T0320397

The Physics of Flight

The Physics of Flight provides a comprehensive explanatory reference on the basic physics of flight with a clear presentation of the underlying mathematics. It presents a momentum-based explanation of lift making no use of Bernoulli's theorem.

Misconceptions are disproved, such as identifying centrifugal force experienced in an airplane undergoing maneuvers as a fictitious force, and not attributing weightlessness during airplane pitch over or experienced in an airplane performing a parabolic flight path to the effects of free fall. This book places particular emphasis on Newton's second law of motion to explain the effects of forces acting on an airplane, the mechanism of lift, and the principles of propulsion.

This book is intended for undergraduate aviation and aerospace students taking courses in Flight Dynamics, Introduction to Flight, and Physics of Flight.

The Physics of Flight
A Primer

Albert Storace, P.E.

CRC Press
Taylor & Francis Group
Boca Raton London New York

CRC Press is an imprint of the
Taylor & Francis Group, an **informa** business

Cover image: Shutterstock

First edition published 2024
by CRC Press
2385 NW Executive Center Drive, Suite 320, Boca Raton FL 33431

and by CRC Press
4 Park Square, Milton Park, Abingdon, Oxon, OX14 4RN

CRC Press is an imprint of Taylor & Francis Group, LLC

ISBN: 978-1-032-48815-8 (hbk)
ISBN: 978-1-032-48816-5 (pbk)
ISBN: 978-1-003-39091-6 (ebk)

DOI: 10.1201/9781003390916

Typeset in Times
by codeMantra

Access the Solutions Manual: routledge.com/9781032488158

To

My wife Jane

and

My daughters Mary, Greta, and Juliette

and

My grandchildren Quinn and Kimberly

Contents

Preface

In this monograph on the physics of flight, it is explained why Newtonian mechanics is the most important tool for understanding flight. Special emphasis is placed on Newton's second law of motion which states that the net force applied to an object is equal to the acceleration of the object times the mass of the object. Newton's second law of motion is the explicit form of the conservation of momentum. The effects of airplane maneuvers on the forces acting on an airplane, the mechanism of aerodynamic lift, the principles of airplane engine dynamics, and of propulsion are all explained by Newton's laws of motion, and focus is placed on describing the physics of aerobatic flight. A conservation of momentum-based explanation of aerodynamic lift with no use made of Bernoulli's theorem is also presented.

Because of the total dependence of flight on Newtonian mechanics, a review is provided in great detail both in inertial and accelerating frames of reference. Misconceptions are disproved such as identifying centrifugal force experienced in an airplane undergoing maneuvers as a fictitious force, not attributing weightlessness during airplane pitch over or experienced in an airplane performing a parabolic flight path to the effects of free fall, and erroneously stating that a barrel roll is a $1\,g$ maneuver.

This monograph is intended primarily for those who wish to broaden their knowledge of the physics of flight. It provides a comprehensive explanatory reference on the basic physics of flight filling the void of an adequate physical and analytical description of this science in the literature. Provided is a detailed explanation of the mathematical modeling and simulation of the physics of flight.

Assuming a first-year college background in physics and math, the material presented guides you through a description of the theory and calculations of the physics of flight.

The author is a licensed professional engineer, jet engine dynamics engineer, a commercial instrumented rated pilot, and an aircraft owner.

Author

Albert Storace, BME, MME, PE

Albert Storace has more than 40 years of industrial experience in the areas of jet engine dynamics, structures, and mechanical design, and has authored numerous design practices, papers, and reports.

He has developed design process improvements and computer codes for jet engine rotordynamics, squeeze-film dampers, and composite structures, and has performed conceptual and detailed design and analysis for Wankel rotating combustion engines and for advanced jet and turbine engines. Albert Storace developed the architecture and design for wireless jet engine radio controlled auto balancing systems.

He developed and coded the VISTA computer program used for the dynamics architecture and rotordynamics design and analysis of all GE Aerospace Jet and aero derivative marine and stationary engines. He developed the Modal Stability Criterion used at GE Aerospace for the prediction of engine dynamic instability caused by blade tip-clearance aerodynamic forces.

In an industrial setting, he has instructed in courses on rotordynamics, jet engine systems integration, and digital signal analysis. Albert Storace managed Air Force programs on jet engine exoskeletal structures and foreign object damage, an Army program on jet engine advanced magnetic bearings controls, and NAVY programs on jet engine magnetic bearings and integral starter-generator systems.

He developed and coded a jet engine transient dynamics analysis program for NASA. Albert Storace performed the design and analysis for a NAVY jet engine aero derivative marine engine shock and vibration mounts that attenuate underwater shock loads into the engine to acceptable levels and minimize the transmission of structure borne noise.

A member of ASME, he has served as a rotordynamics session organizer for various ASME International Gas Turbine Institute (IGTI) meetings.

Albert Storace has received the best paper awards from the ASME Structures and Dynamics and the Turbomachinery Committees including the Melville Medal, the highest honor for the best original technical paper published in the ASME Transactions over a 2-year period.

Albert Storace earned his bachelor's and master's degrees in mechanical engineering at the City College of New York. He is a licensed professional mechanical engineer in Ohio and holds five patents.

He holds the commercial pilot license with instrument rating and is an aircraft owner.

1 An Overview of Newton's Laws of Motion

Newton's laws of motion only hold their usual form in an inertial frame, that is, a frame that is neither accelerating nor rotating. However, a reference frame fixed to the earth is an excellent approximation of an inertial frame, even though the earth is rotating on its axis and accelerating in its orbit.

1.1 NEWTON'S FIRST LAW

In an inertial frame of reference, an object either remains at rest or continues to move at a constant velocity in a straight line, unless acted upon by a force. This is the law of inertia.

1.2 NEWTON'S SECOND LAW

In an inertial reference frame, the vector sum of the forces F on an object is equal to the mass m of that object multiplied by the acceleration a of the object:

$$F = m\frac{d^2 r_o}{dt^2} = m\ddot{r}_o = ma \tag{1.1}$$

where ma is not a force and r_o is the position of the object relative to the inertial frame. This is the standard form of Newton's second law.

$F = ma$ is the explicit form of the conservation of momentum and is also defined as the equation of motion, a differential equation of second order (it is assumed here that the mass m is constant).

However, in an accelerating frame of reference, the form of Newton's second law (the equation of motion) is different as there is an extra force term, the inertia force $-mA$, as explained by Taylor (Ref. [1]), such that

$$F - mA = m\frac{d^2 r}{dt^2} = m\ddot{r} \tag{1.2}$$

where A is the acceleration of the moving frame relative to the inertial frame, r is the position of the object relative to the moving frame, and $\ddot{r}$ is the acceleration of the object relative to the moving frame. The development of equation 1.2 is presented in Chapter 2.0.

The inertial force mA is sometimes called a fictitious force, but it is a real force in the accelerating frame of reference. Taking a free body diagram of the mass, the actual forces acting are F and $-mA$. Also note that by definition from the second law, $-mA$ is a force.

DOI: 10.1201/9781003390916-1

Note that it is said in the literature that Newton's second law (actually the standard form of the second law) is valid only in inertial frames, that is, frames that are neither accelerating nor rotating. However, the equation of motion developed from Newton's second law in the accelerating reference frame is $m\ddot{r} = F - mA$. This equation is a mathematical representation of what is meant by the statement that Newton's second law (the standard form) does not apply in non-inertial frames. It is not that the physics dealing with Newtonian mechanics cannot be analyzed in a non-inertial frame, but that the formulation of the equations of motion is different.

1.3   NEWTON'S THIRD LAW

In inertial or accelerating frames of reference, when one body exerts a force on a second body, the second body simultaneously exerts a force equal in magnitude and opposite in direction on the first body.

> **Comments:** Note that the first and third laws are rather obvious and intuitive and require no additional discussion.

PROBLEMS

1. Describe Newton's second law of motion in an inertial frame and with respect to an accelerating frame.
2. Describe why the inertia force with respect to an accelerating frame is a real force.

2 Newton's Second Law of Motion in Non-Inertial Frames with Translational Acceleration

Consider an inertial frame Δ_0 and a second frame Δ that is accelerating relative to Δ_0 with an acceleration A. Refer to Figure 2.1. The second frame Δ has a velocity V and an acceleration $A = \dot{V}$ relative to the inertial frame Δ_0. Because Δ_0 is inertial, Newton's second law holds and hence $m\ddot{r}_0 = F$ where r_0 is an object's position relative to Δ_0 and F is the resultant or net force (gravity force, air resistance, friction force, spring forces, etc.) acting on the object. Now consider the object's motion measured relative to the accelerating frame Δ. The object's position relative to Δ is r and by vector addition of velocities, its velocity $\dot{r}_0$ relative to the inertial frame Δ_0 is $\dot{r}_0 = \dot{r} + V$, that is, the object's velocity relative to the inertial frame equals the object's velocity relative to the accelerating frame plus the accelerating frame's velocity relative to the inertial frame. Then, $\ddot{r}_0 = \ddot{r} + A$ and $\ddot{r} = \ddot{r}_0 - A$. Thus the equation of motion, equation 1.2, in the accelerating frame Δ is $m\ddot{r} = m\ddot{r}_0 - mA = F - mA$ where m is the object's mass and F is the resultant force acting on the object exclusive of the inertia force $-mA$. This equation of motion has exactly the form of Newton's second law in an inertial frame, except that in addition to F, the sum of all the forces, there is an extra force on the right side equal to $-mA$ termed the inertia force.

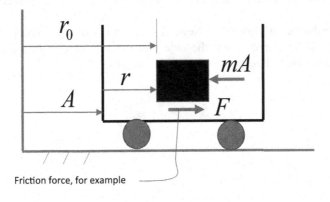

FIGURE 2.1 Accelerating reference frame.

DOI: 10.1201/9781003390916-2

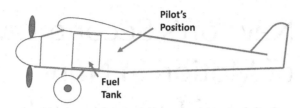

FIGURE 2.2   Forward located fuel tank in the Spirit of St. Louis airplane.

The equation of motion in the accelerating reference frame is $m\ddot{r} = F - mA$ where A is the acceleration of the accelerating frame relative to the inertial frame. To an observer in the accelerating frame, the mass would appear to move under the action of the entire force on the right hand side which is the sum of all the forces equal to F plus the inertia force $-mA$. Note that the inertia force $-mA$ is sometimes erroneously called a fictitious force. The term fictitious force arises from the fact that this force will not appear in an inertial frame but is very much a real force in the non-inertial frame.

When the mass is at rest in the accelerating frame, that is, its acceleration $\ddot{r}$ relative to the accelerating frame is zero, then $m\ddot{r} = 0 = F - mA$ and thus $F = mA$.

It will be noted that the fuselage fuel tank in Lindbergh's plane, the Spirit of Saint Louis, was located forward to the pilot's seat as shown in Figure 2.2 to minimize a change in the location of the center of gravity as fuel was consumed and to prevent the pilot from being crushed in the event of crash induced rear acting acceleration (actually a deceleration or a reduction of the forward velocity) of the plane.

This acceleration causes a forward acting inertia force that would move the gas tank away from the pilot in the forward direction in the event of a crash. This is an example of the force system in a non-inertial frame (the airplane).

PROBLEMS

1. Describe the forces acting on an object with respect to an accelerating frame.
2. Describe the equivalence of Newton's second law of motion in inertial and accelerating frames when the object in the accelerating frame has zero acceleration with respect to the accelerating frame.

3 Newton's Second Law of Motion in Inertial and Non-Inertial Frames with Acceleration due to Rotation

In an inertial reference frame, the centripetal force acting on a body traveling in a circular path results in centripetal acceleration (inward radial acceleration). The centripetal force direction is always orthogonal to the motion of a body and toward the fixed point of the instantaneous center of curvature of the path. In Newtonian mechanics, the force of gravity provides the centripetal force responsible for astronomical orbits. One common example involving centripetal force is the case in which a body moves with uniform speed along a circular path. The centripetal force is directed at right angles to the motion and also along the radius toward the center of the circular path.

Referring to Figure 3.1, mass m is moving in a circular path with tangential velocity V and angular velocity ω relative to the inertial frame. The accelerating centripetal force F causes the body to follow a curved path and results in centripetal acceleration $A = \dfrac{V^2}{r} = r\omega^2$. Newton's second law in an inertial frame provides $F = mA = m\dfrac{V^2}{r}$ and mA is not a force.

If the mass m was constrained by a string fixed at the center of curvature, then the string would be in tension with force F.

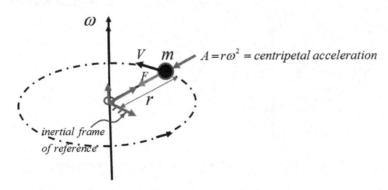

FIGURE 3.1 Body moving in a circular path under the influence of centripetal force F.

DOI: 10.1201/9781003390916-3

$$F = mA = m\frac{V^2}{r} = mr\omega^2$$

Taking a free body diagram of the mass, the only force acting on the mass is F and $mA = m\dfrac{V^2}{r}$ is not a force.

Now if the mass m is at rest in a rotating frame of reference rotating with angular velocity ω relative to the inertial frame as shown in Figure 3.2, then from similarity to the previously discussed translational frame mechanics and noting that the amplitude of the acceleration A of any point in the rotating frame at radius r relative to the inertial frame is $\dfrac{V^2}{r} = r\omega^2$ and is pointed inward, then the equation of motion relative to the rotating frame is $0 = m\ddot{r} = F - mA = F - m\dfrac{V^2}{r}$ where $\ddot{r}$ is the acceleration of m relative to the rotating frame and $-mA$ is the outward acting centrifugal force. This equation is the same as equation 1.2. It is assumed that the mass has no velocity relative to the rotating frame, that is, the Coriolis force which acts perpendicular to the velocity direction and the angular velocity axis is equal to zero and thus $\ddot{r} = 0$.

This is Newton's second law of motion in a non-inertial frame expanded to include the acceleration of the non-inertial frame relative to the inertial frame.

Again, $m\ddot{r} = 0 = F - mA$ and thus $F = mA$.

For example, this could be a mass resting on a turn table and F is a constraint force provided in the rotating frame (could be a friction force).

Figure 3.2 shows the mass resting in the rotating frame of reference. Taking a free body diagram of the mass, the forces acting on the mass are F and $-mA$.

On comparing the force systems for the inertial and rotating frames shown, respectively, in Figures 3.1 and 3.2, it is seen in the inertial frame that $F = mA = m\dfrac{V^2}{r} = mr\omega^2$

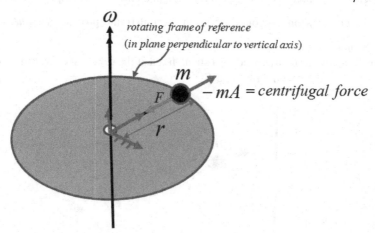

FIGURE 3.2　Forces acting on mass located in a rotating frame of reference.

and in the rotating frame, $F = mA = m\dfrac{V^2}{r} = mr\omega^2$. In the former case, F is the accelerating force that results in the acceleration A and in the latter case, F is the restraint force that balances the centrifugal force mA. In the inertial frame, mA is not a force and in the rotating frame, the centrifugal force mA is a real force. This centrifugal force is an inertial force and is sometimes erroneously called a fictitious force because it is equal to mA which is not a force in an inertial reference frame. In summary, the equilibrium equation $F - mA = 0$ is the same for both the inertial and rotating frames.

It is very much a misnomer to call the centrifugal (inertia) force which may black out the pilot of an aircraft "fictitious". As another example, consider a projectile with no obvious force acting on it. It does not remain at rest or move uniformly in a straight line but is accelerated downward by the force of gravity which is the weight of the projectile. No one can object to gravitational forces or has suggested that they can be termed "fictitious".

An example of a system that subjects an individual to a centrifugal force in a rotating frame is the human-rated centrifuge. The United States Air Force centrifuge located at Wright Patterson Air Force base and shown in Figure 3.3 allows individuals (seated in a gondola at the left end of the rotating arm in Figure 3.3) to experience up to $20\,g$, or 20 times the normal force of gravity in a variety of directions, to teach the effects of G forces on human physiology and to measure the subject's ability to counteract the effects and prevent G-induced loss of consciousness. This centrifuge has an arm of 31 ft and the multi axis control of the gondola allows positive-g, with the centrifugal force acting in a head to foot direction, and negative-g, with the centrifugal force acting in the foot to head direction, and also allows the centrifugal force to act in the transverse chest-to-back direction.

FIGURE 3.3 The United States Air Force centrifuge at Wright Patterson Air Force Base.

PROBLEMS

1. Describe Newton's second law of motion with respect to a rotating frame.
2. Describe why in a rotating frame the centrifugal force is a real force.

4 G Loading and Weightlessness

The acceleration of a freely falling body is called the acceleration due to gravity or the acceleration of gravity and is denoted by the letter g. At or near the earth's surface, the value of g is approximately by 32ft/sec^2 (9.8 m/sec^2). Note that the quantity "g" is sometimes referred to simply as "gravity" or as "the force of gravity" both of which are incorrect. The force of gravity means the force with which the earth attracts a body otherwise known as the weight W of the body and $W = mg$ (pounds in the English system and Newton in the cgs system) where m is the mass of the body. The letter "g" represents the acceleration caused by the force of gravity.

G force or G equals nW where the load factor n is a multiplier of the body's weight. Hence, 2G represents a force acting on the body equal to two times the body's weight or a force that causes an acceleration equal to $2 \times g$.

The following is a discussion of what is meant by the term weightlessness. Weightlessness for a falling object occurs when its acceleration equals the acceleration of gravity and thus it is in free fall. The only force acting on the object is therefore the force of gravity or the weight of the object. Figure 4.1 shows an example of weightlessness consisting of a freely falling elevator where friction and aerodynamic forces acting on the elevator are neglected. The elevator and its contents are weightless and in free fall.

In the inertial frame, from Newton's second law, the equation of motion for an object in the elevator is:

$$F - W = F - mg = -mg$$
$$F = mg - mg = 0$$

Gravitational
Force

Thus, there is no reaction force acting on the object. The acceleration of the object is equal to the acceleration of gravity and it is weightless and in free fall.

Another example of weightlessness is provided by the so-called reduced gravity aircraft (sometimes called the vomit comet). This aircraft gives its occupants the sensation of weightlessness by following a parabolic flight path. This flight profile is chosen to provide a relatively gentle means of introducing downward acting radial acceleration without pitching the airplane over and entering an initial dive. While following the parabolic path which occurs in the upper zone of the maneuver as the aircraft arches over in a ballistic trajectory and behaves as a projectile with engine thrust equal to drag and in a zero-lift configuration, the only force acting on the aircraft and its payload is the gravity force $W = mg$ as shown in Figure 4.2 and thus the aircraft is in free fall and therefore weightless. From Newton's second law, equating the gravitational force to the centripetal acceleration $\times$ mass,

DOI: 10.1201/9781003390916-4

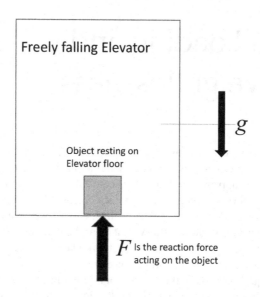

FIGURE 4.1   Freely falling elevator.

FIGURE 4.2   Inertial frame force and acceleration acting on the aircraft at the top of the maneuver.

$$W = mg = m\frac{v^2}{r_p} \text{ and thus, } g = \frac{v^2}{r_p}$$

Thus the acceleration g due to the gravity force equals the centripetal acceleration $\dfrac{v^2}{r_p}$ where v and r_p are the tangential velocity and the instantaneous radius of curvature, respectively.

Note from equation 1.2, as a consequence of free fall, the payload relative to the aircraft is subjected to the centrifugal force resulting from the downward acceleration of gravity in the inertial frame. Thus, in the upper zone of the maneuver, the payload gravity force is balanced by the centrifugal force resulting in net zero force acting on the payload relative to the aircraft frame of reference. Therefore, the payload is weightless and has zero acceleration relative to the aircraft. Note that there is the misconception that in an inertial frame, the reduced gravity aircraft provides weightlessness because the centrifugal force is balanced by the gravity force.

To summarize what weightlessness is:

When the object is in free fall, its acceleration is equal to the acceleration of gravity as the only force acting on the object is the gravity force.

PROBLEMS

1. Describe what free fall is and why it results in a weightless condition.
2. Develop the equation that defines the aircraft velocity needed to achieve free fall at the top of the maneuver used for the so-called reduced gravity aircraft.
3. What are the forces acting on the payload of the so-called reduced gravity aircraft when it is in free fall?

5 Aerodynamic Lift

5.1 AERODYNAMIC LIFT FUNDAMENTAL CONCEPTS

5.1.1 THE POPULAR EXPLANATION OF LIFT DEVELOPMENT

Bernoulli's energy equation is the basis for the popular explanation of how aerodynamic lift is developed. Bernoulli's equation states that the total mechanical energy of an incompressible moving fluid comprising the gravitational potential energy of elevation, the energy associated with the fluid static pressure, and the kinetic energy of the fluid motion remains constant.

The explanation of how lift is developed using Bernoulli's equation is the fact that larger velocities imply lower pressures and thus a net upwards pressure force is generated as the air flows around the airfoil because it was theorized that the airflow at the top of the airfoil has a higher velocity than the airflow at the bottom of the airfoil. The airflow is characterized by streamlines which define the path of a particle of air relative to solid body, the airfoil as shown in Figure 5.1.1. A fallacy in this explanation is that the airflow velocity is higher at the top of the airfoil because of equal transit time. The principle of equal transit time states that the airflow separates at the leading edge of the airfoil and must rejoin at the trailing edge (points a and b, shown, respectively, in Figure 5.1.1) at the same time.

Although Figure 5.1.1 shows a symmetrical airfoil shape, the equal transit time principle assumes a typical asymmetrical airfoil where there is a slightly longer distance that the airflow must travel over the upper surface compared to the lower surface. Thus the airflow over the upper surface has a longer distance to travel and must therefore go faster to reach the trailing edge at the same time as the airflow over the lower surface of the airfoil. However, there is no physical reason which indicates that the airflow from above and below the airfoil must meet up at the trailing edge at the same time. The fallacy of the equal transit time principle is shown from the results of smoke wind-tunnel experiments reported in Reference [12]. Also note that it would be impossible for an airplane to fly inverted if the equal time theory were true. Based on the lack of real physics and experimental evidence, the equal time theory is clearly incorrect.

5.1.2 AN ALTERNATIVE EXPLANATION OF LIFT BASED ON AIRFLOW CURVATURE

Presented is a description of how airflow curvature develops aerodynamic lift. Lift is explained herein by the action of pressure gradients caused by the curvature of streamlines. It is hypostatized that this is the most significant factor in the generation of the lift force. Consider the conservation of momentum explicitly expressed by Newton's second law of motion. Normal to the streamline accelerations in the fluid field sustained by normal pressure forces acting above and below the airfoil surfaces result in the lift force. Thus any shape or configuration that introduces curvature

DOI: 10.1201/9781003390916-5

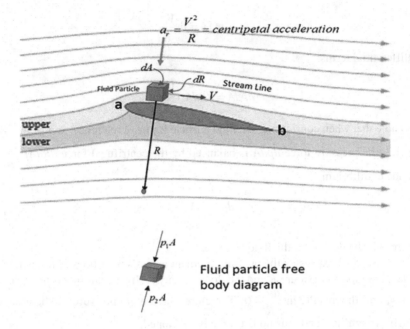

FIGURE 5.1.1 Streamlines flowing around an airfoil, the depiction of an air fluid particle, the pressures, the centripetal acceleration, and the free body diagram.

into the flow field can generate lift. Airfoils work because the flow follows the local surface curvature on the upper and lower surfaces.

As the air flows over the airfoil surfaces, its path is curved by normal pressure forces that drive centripetal acceleration. Above the airfoil upper surface, the pressure increases in the normal direction away from the airfoil surface until the pressure reaches the ambient pressure. Thus, a pressure gradient is created, where the pressure on the upper airfoil surface is less than the ambient pressure. Below the airfoil lower surface, the pressure decreases in a normal direction away from the airfoil until the pressure reaches the ambient pressure such that there is a pressure higher than the ambient pressure at the airfoil lower surface. The difference in airfoil upper and lower surface pressure forces results in a net upward acting lift force.

It will be noted that radial direction conservation of momentum-based explanations of lift is also discussed by Anderson and Eberhardt (Ref. [2]), McLean (Ref. [3]), Babinsky (Ref. [12]), and Gregory (Ref. [14]).

Refer to Figure 5.1.1 for a graphical description of the streamlines flowing around an airfoil, the depiction of an air fluid particle, the pressures, the centripetal acceleration, and free body diagram.

Note again that the pressure gradient across the streamline is caused by the curved streamline, with pressure increasing in the direction away from the center of curvature.

From the inertial frame free body diagram in Figure 5.1.1 and equating the sum of the forces to the radial acceleration (centripetal acceleration) times mass where m is the mass of the fluid particle (Newton's second law):

$$(p_1 - p_2)A = m\frac{V^2}{R}$$

In differential terms:

$$dpdA = dRdA\rho\frac{V^2}{R}$$

Also note that when considering the fluid particle frame of reference, and from equation 1.2, the pressure force $dpdA$ is balanced by the centrifugal force $dRdA\rho\frac{V^2}{R}$ to provide equilibrium.

$$\frac{dp}{dR} = \rho\frac{V^2}{R} \tag{5.1.1}$$

Where ρ is the density of the fluid particle.

Equation 5.1.1 expresses the pressure gradient across streamlines as a function of the local radius of curvature R and the flow velocity V. If a streamline is not turning, $R \rightarrow \infty$, and this results in $\frac{dp}{dR} = 0$. Therefore, there is no pressure gradient across straight streamlines and thus no lift force is developed.

5.2   AERODYNAMIC LIFT EQUATION

The lift coefficient C_L is a dimensionless coefficient that relates the lift force generated by a lifting body such as an airfoil to the fluid velocity, the fluid density, and an associated reference area. It is a function of the angle of attack α (Figure 5.2.1) of the body, the angle between the chord line and the relative wind. The lift coefficient can be calculated using thin airfoil theory, calculated numerically, or determined from wind-tunnel tests. Plots of C_L versus the angle of attack show the same general shape for all airfoils, but the particular values will vary, that is, C_L is specific to a particular airfoil shape. Figure 5.2.2 shows an example plot. The lift force is one of the force components of the total aerodynamic force. It is perpendicular to the relative wind. The other component is the induced drag force which is parallel to the relative wind. Refer to References [4] and [5] for further discussion.

The lift coefficient C_L is defined by:

$$C_L = \frac{F_L}{qA_r} = \frac{F_L}{\frac{1}{2}\rho V_R^2 A_r} = \frac{2F_L}{\rho V_R^2 A_r} \tag{5.2.1}$$

where F_L is the lift force

q is the fluid dynamic pressure
V_R is the relative wind velocity (true airspeed)
ρ is the fluid density
A_r is the wing surface area

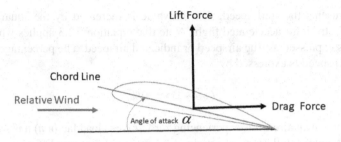

FIGURE 5.2.1 Angle of attack.

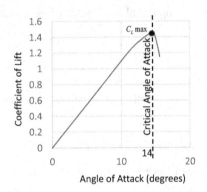

FIGURE 5.2.2 Lift coefficient versus angle of attack.

Note that the coefficient of lift increases steadily with increasing angle of attack until C_L max and then begins to drop as the wing stalls. For a given airfoil, a stall always occurs at the same angle of attack, which is approximately 10°–15° for a typical airfoil.

Solving equation 5.2.1 for the lift force F_L, the lift equation (equation 5.2.2) is obtained.

$$F_L = \frac{1}{2} C_L \rho V_R^2 A_r \qquad (5.2.2)$$

Note that in straight and level flight, C_L max is normally reached as the indicated airspeed approaches the published stall speed, that is, as airspeed decreases, the angle of attack must increase to maintain the lift needed for level flight. Also note that stall speed increases as g-loading increases during maneuvers such as level turns because the load factor n increases the effective weight of the aircraft. Thus the increase in effective weight must be balanced by a similar increase in lift of the wings, necessitating a higher angle of attack.

Rearranging the lift equation to solve for the velocity, we obtain equation 5.2.3 where the lift force has been replaced with the product of the aircraft's weight W and the load factor n. The load factor n is discussed in chapter 4.

$$V_R = \sqrt{\frac{2F_L}{C_L \rho A_r}} = \sqrt{\frac{2nW}{C_L \rho A_r}} = \sqrt{n}\sqrt{\frac{2W}{C_L \rho A_r}} \qquad (5.2.3)$$

If V_R represents the stall speed, then its value is increased by the square root of the load factor n for accelerated flight. Note that equation 5.2.3 applies whether the velocity is expressed as true airspeed or indicated airspeed. The percentage increase in the stall speed is expressed by

$$(\sqrt{n} - 1) \times 100\% \qquad\qquad (5.2.4)$$

Note that when an aircraft is experiencing zero gs (zero load factor n) it is weightless, and thus it cannot stall because the wings are not generating any lift, that is, they are unloaded.

For example, the inside loop radial load factor (from equation 6.4.1) at the top of the loop for zero gs is:

$$n = \frac{v^2}{gr} - 1 = 0$$

where the number of gs for the centrifugal acceleration is equal and opposite to the gravitational force g.

PROBLEMS

1. Explain why the use of Bernoulli's theorem to describe how the lift force is created is clearly incorrect.
2. Describe the alternative explanation of lift based on airflow curvature.
3. Describe the effect of angle of attack on the lift force and on the stall speed.
4. Describe the effect of the load factor on the stall speed.

6 Airplane Maneuvers

6.1 THE TURN

The turn is accomplished by a banking (rolling) maneuver that changes the heading of the aircraft. The turn is initiated by using the ailerons to roll the aircraft. Figure 6.1.1 shows an aircraft rolling to the left by lowering the right aileron and raising the left aileron. This effectively increases the curvature (camber) of the right wing thereby increasing its lift force and effectively decreases the curvature of the left wing thereby decreasing its lift force. In a coordinated turn (the aircraft's longitudinal axis is perpendicular to the turn radius and thus aligned with the relative wind), the net wing lift aerodynamic force vector **L** is directed perpendicular to the wings generating the lift.

As the aircraft is rolled, the lift force vector is tilted in the direction of the roll. The vertical lift force component is opposed by the weight vector. The horizontal lift force is unopposed. The free body diagram of the aircraft performing a level coordinated turn illustrating the forces acting is shown in Figure 6.1.1.

As long as the aircraft is banked, the side force (the horizontal component of the lift force) is a constant and unopposed force on the aircraft. This force is the net inward accelerating force acting on the aircraft and causes centripetal acceleration a_r of the aircraft. The resulting motion of the center of gravity of the aircraft is a circular arc. Note that the rudder is not used to turn the aircraft and that the aircraft is turned by the action of the side component of the lift force. However, the rudder is used to correct any deviation between the longitudinal axis of the airplane and the tangent to the turn circle, that is, the rudder is used rolling into the turn to bring the nose back in line with the turn circle tangent. This deviation is caused by the

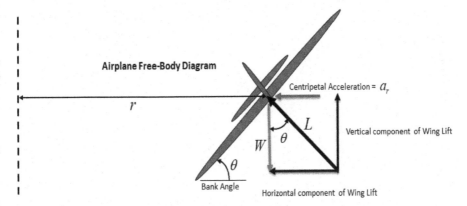

FIGURE 6.1.1 Forces acting on the banked aircraft undergoing a level coordinated turn shown in an inertial reference frame.

DOI: 10.1201/9781003390916-6

adverse yaw moment resulting from the increased induced drag force on the outer wing caused by the lowered aileron. If no rudder is used in a turn, the adverse yaw moment can cause the nose of the aircraft to yaw to the outside of the turn.

For a level coordinated turn at a bank angle of θ radians:

$$L\cos\theta = W$$

$$L = (W/\cos\theta)\,\text{lbs} \qquad = \text{Vector sum of the aircraft weight and}$$
$$\text{the horizontal component of lift}$$

For 60 degree bank,

$$L = W/.5 = 2W$$

$$n = \frac{L}{W} = \frac{1}{\cos\theta} = \text{Load Factor(number of } gs\text{)}, \qquad (6.1.1)$$

where θ is the bank angle in radians.

Figure 6.1.2 shows this load factor plotted versus the bank angle.

Note that in a level turn the lift force must be increased so that its vertical component equals the weight of the airplane. This requires that the elevator be used to increase the airplane pitch and thus to increase the wing angle of attack. Thus, while the ailerons act as the primary turn control, the elevator is required to maintain a level turn. Note that in Reference [10], it is stated that the elevator is the primary turn control and clearly this is not the case.

Referring to Figure 6.1.1, the radius of the turn for a coordinated turn is obtained from Newton's second law in an inertial frame as follows:

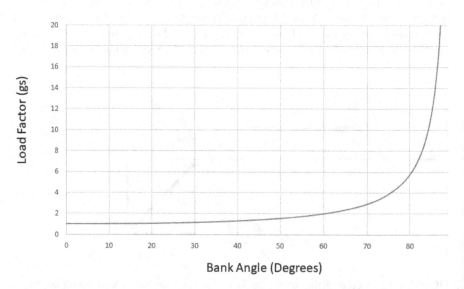

FIGURE 6.1.2   Load factor plotted versus the bank angle for a level coordinated turn.

$$ma_r = \frac{W}{g}r\omega^2 = \frac{W}{g}\frac{v^2}{r} = HCL = L\sin\theta = W\tan\theta$$

where HCL is the horizontal component of the lift force, v is the aircraft tangential velocity, and r is the radius of the turn.

Solving for r:

$$r = \frac{v^2}{g\tan\theta} = \frac{(v=1.6878\frac{\text{ft/sec}}{\text{knots}}\times\text{knots})^2}{32.174\times\tan\theta} = \frac{.0884\times\text{knots}^2}{\tan\theta} =$$

(6.1.2)

$$= \frac{\text{knots}^2}{11.29\tan\theta}\,\text{ft} = \left(.02699\frac{\text{knots}^2}{\tan\theta}\right)\text{meters}$$

Thus, an increase in airspeed results in an increase of the turn radius for a given bank angle.

The rate of turn (angular velocity) is also obtained from Newton's second law as follows:

$$\omega = \sqrt{\frac{g}{r}\tan\theta} = \sqrt{g\tan\theta\frac{1}{r}} = \sqrt{g\tan\theta\frac{g}{v^2}\tan\theta} = \sqrt{\frac{g^2}{v^2}\tan^2\theta}\;\frac{rad}{sec}$$

$$\omega = \sqrt{\frac{g^2}{v^2}\tan^2\theta} = \frac{g}{v}\tan\theta = \frac{32.174\tan\theta}{v=1.6878\frac{\text{ft/sec}}{\text{knots}}\times\text{knots}} =$$

$$= \frac{19.0626\tan\theta}{\text{knots}}\;\frac{rad}{sec}$$

Rate of turn (ROT) =

$$= \frac{19.0626\tan\theta}{\text{knots}}\;\frac{rad}{sec}\times57.2956\,\text{deg/rad} =$$

(6.1.3)

$$= \frac{1{,}092.2\tan\theta}{\text{knots}}\;\text{deg/sec}$$

Thus, an increase in airspeed results in a decrease in the rate of turn for a given bank angle.

The bank angle for a given rate of turn is obtained from:

$$\omega = \frac{g}{v}\tan\theta\;\frac{rad}{sec}$$

$$\theta = \tan^{-1}(\frac{\omega v}{g})\;\text{rad}$$

$$\text{Bank Angle in degrees} = \tan^{-1}\left(\frac{\text{ROT} \times \text{KNOTS}}{1,092.2}\right) \qquad (6.1.4)$$

A standard rate of turn for a light training aircraft is 3 deg/sec. Thus, 2 minutes is required to turn 360°.

At an airspeed of 100 knots, the required bank angle for the standard rate of turn is obtained as follows:

$$\text{Bank Angle} = \tan^{-1}\left(\frac{3 \times 100}{1,092.2}\right) = 15.358°$$

Therefore, for approximately 15° of bank at an airspeed of 100 knots, the 2 minutes turn coordinator indicator points to the reference mark thus indicating a standard rate of turn.

Now consider Newton's second law of motion in the aircraft reference frame which is a non-inertial frame. In a turn, the aircraft is undergoing centripetal acceleration $\frac{v^2}{r} = r\omega^2$ and the equation of motion relative to the aircraft (as previously mentioned for a rotating frame and as shown by equation 1.2) is $m\ddot{r} = F - mA = F - m\frac{v^2}{r}$ where $\ddot{r}$ is the acceleration of a mass m located in the aircraft which is assumed to be zero and $-mA = -m\frac{v^2}{r}$ is the centrifugal force acting on the mass m and F is the sum of the other forces acting on the mass m. Consider the free body diagram of the pilot shown in Figure 6.1.3. Force equilibrium in the horizontal direction gives

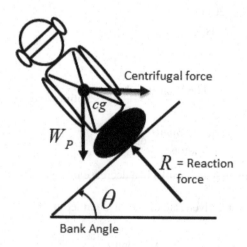

Pilot Free-Body Diagram

FIGURE 6.1.3 Pilot free body diagram for an aircraft coordinated turn in an aircraft reference system.

$$-R\sin\theta + \frac{W_P}{g}\frac{v^2}{r} = 0$$

This shows that centrifugal force actually acts on the pilot in a turning aircraft and R is the reaction force exerted on the pilot by the seat.

Force equilibrium in the vertical direction gives

$$R\cos\theta = W_P \text{ or } R = W_P/\cos\theta$$

and thus, the load factor for a coordinated turn is

$$n = \frac{R}{W_P} = \frac{1}{\cos\theta}.$$

For a level coordinated turn at a 60° bank, the load factor is 2.0 and thus a reaction force equal to two times the pilot's weight is acting on the pilot. The load factor is the amplitude of the vector sum of the pilot's weight and the centrifugal force divided by the pilot's weight.

In the aircraft reference system, the forces acting on the aircraft in coordinated and uncoordinated turns and the corresponding indications of the slip/skid indicator will be discussed.

The slip/skid indicator is also known as an inclinometer, and it indicates the quality of the turn. It is mounted in the bottom of the turn-bank indicator or the turn coordinator. The inclinometer consists of a metal ball in an oil-filled, curved glass tube and the ball is loaded by centrifugal and gravitational forces. When the aircraft is performing a coordinated turn, the ball remains centered at the bottom of the glass tube.

Figure 6.1.4 shows the forces acting on the aircraft during a level coordinated turn. Note that the aircraft's resultant aerodynamic (Lift) force lies in the plane of

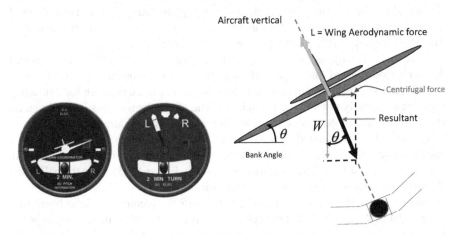

FIGURE 6.1.4 Inclinometer at the bottom of the turn coordinator or the turn and slip indicator, inclinometer indication, and forces acting on the aircraft (in an aircraft reference system) during a level coordinated turn.

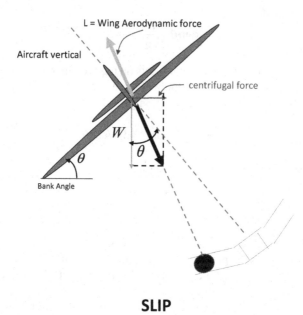

SLIP

FIGURE 6.1.5 Inclinometer indication, and forces acting on the aircraft (in an aircraft reference system) during a level uncoordinated turn with slip.

symmetry perpendicular to the aircraft's lateral axis for a coordinated turn. For this condition, the ball in the slip/turn indicator is centered at the bottom of the inclinometer. Also note that the forces acting on the ball are the ball's weight and the centrifugal force acting on the ball.

For the coordinated or uncoordinated turn, there must be equilibrium between the horizontal lift force component and the centrifugal force. This is obvious since the horizontal lift force component accelerates the airplane radially toward the center of the turn resulting in a centrifugal force that is equal and opposite to the horizontal lift force component in the airplane frame of reference.

Note that in many references, for example, References [6] and [17], it is stated that for uncoordinated turns, the horizontal lift force component is greater than the centrifugal force for a slipping turn and less than the centrifugal force for a skidding turn. This is clearly not a correct statement. The reason behind this incorrect statement is explained in the discussion below for slipping and skidding turns.

For a level uncoordinated turn where the aircraft is slipping, the nose of the aircraft is yawed toward the outside of the turn because there is insufficient rudder in the direction of the turn. This condition is shown in Figure 6.1.5 where the ball in the slip/turn indicator falls to the inside of the turn.

The incorrect statement cited above relative to the inequality of the horizontal component of the lift force and the centrifugal force is based on incorrectly expressing the horizontal component of the lift force in terms of the bank angle θ for the slipped condition. This provides a value for the horizontal component of the lift force that is greater than the actual value which is also equal to the centrifugal force.

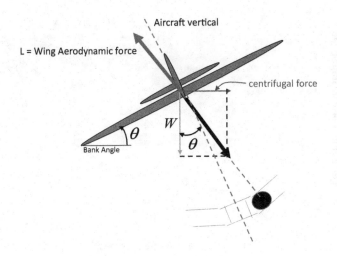

SKID

FIGURE 6.1.6 Inclinometer indication, and forces acting on the aircraft (in an aircraft reference system) during a level uncoordinated turn with skid.

For a level uncoordinated turn where the aircraft is skidding, the nose of the aircraft is yawed toward the inside of the turn because there is excessive rudder in the direction of the turn. This condition is shown in Figure 6.1.6 where the ball in the slip/turn indicator falls to the outside of the turn.

Note that for the skidding turn, incorrectly using the bank angle θ to determine the horizontal component of the lift force provides a value that is smaller than the actual value which is also equal to the centrifugal force.

6.2 BANK ANGLE EFFECT ON STALL SPEED

As shown in Section 5.2, the stall speed increases as g-loading increases during maneuvers such as level turns because the load factor n increases the effective weight of the aircraft, necessitating a higher angle of attack. Substituting the value for the load factor as a function of the bank angle from equation 6.1.1 into equation 5.2.3, we obtain for a level coordinated turn:

$$V_R = \sqrt{n}\sqrt{\frac{2W}{C_L \rho A}} = \frac{1}{\sqrt{\cos\theta}}\sqrt{\frac{2W}{C_L \rho A}}$$

Thus, the percentage increase in the stall speed as a function of bank angle from equation 5.2.4 is given by:

$$(\frac{1}{\sqrt{\cos\theta}} - 1) \times 100\% \tag{6.2.1}$$

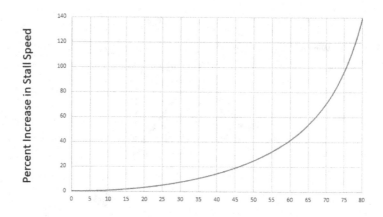

Bank Angle (degrees)

FIGURE 6.2.1   Percent increase in stall speed as a function of bank angle.

Figure 6.2.1 shows a plot of the percentage increase in the stall speed as a function of the bank angle.

Note that the increase in stall speed with the bank angle can be a recipe for disaster for the base to final turn in the traffic pattern if the turn to final is made late thus overshooting the runway. If the pilot tightens the turn by increasing the bank angle in an attempt to realign with the runway, then because of the increased stall speed with increased bank angle, a rarely survivable aerodynamic stall is likely because the airplane is low and slow.

6.3   THE VELOCITY-LOAD FACTOR DIAGRAM

The *V-n* diagram relates airplane airspeed (velocity) to the load factor (*n*). Each *V-n* diagram applies to one airplane type and is valid only for a specific weight, configuration, and altitude. *V-n* diagrams provide the maximum amount of positive or negative lift the airplane is capable of generating at a given speed.

As an example, generate the *V-n* diagram for a typical general aviation aircraft. First calculate the wings level 1*g* stalling speed using the following specifications.

$$n = +1$$

$$\text{Aircraft Weight} = W = 2{,}450 \text{ lbs} = 10{,}898.1439 \text{ Newtons}$$

$$\text{Wing Area} = A_r = 174 \text{ ft}^2 = 16.1651 \text{ m}^2$$

$$\text{At sea level: Air Weight Density} = .07647 \frac{\text{lb}}{\text{ft}^3}$$

$$\text{Air Mass density} = \rho = (.07647 \text{ lb/ft}^3) / (32.174 \text{ ft/sec}^2) = .0023767 \frac{\text{lb-sec}^2}{\text{ft}^4} =$$

$$= .0023767 \frac{\text{slugs}}{\text{ft}^3} = 1.225 \frac{\text{kg}}{\text{m}^3}$$

Use the lift coefficient at the critical angle of attack shown in Figure 5.2.2.

$$C_L = 1.42$$

From equation 5.2.3, the +1g stall speed is:

$$V_S = \sqrt{\frac{2nW}{C_L \rho A_r}} = \sqrt{\frac{(2)(1.0)(2450)}{(1.42)(.0023767)(174)}} = 91.3464 \text{ ft/sec} = 54.121 \text{ knots}$$

and in SI units:

$$V_S = \sqrt{\frac{2nW}{C_L \rho A_r}} = \sqrt{\frac{(2)(1.0)(10898.1439)}{(1.42)(1.225)(16.1651)}} = 27.8414 \text{ m/sec} = 54.121 \text{ knots}$$

From equation 5.2.3, the following second order equations expressing the load factor n in terms of the velocity are obtained.

For positive load factors:

$$n = +\frac{1}{2} \frac{C_L \rho A_r}{W} V^2 \qquad (6.3.1)$$

For negative load factors:

$$n = -\frac{1}{2} \frac{C_{LN} \rho A_r}{W} V^2 \qquad (6.3.2)$$

In equation 6.3.1 for positive load factors, C_L represents the lift coefficient for non-inverted flight and in equation 6.3.2 and C_{LN} represents the lift coefficient for inverted flight. Note that for inverted flight, the resulting inverted airfoil is not as aerodynamically efficient as the airfoil shape for non-inverted flight and thus the value for C_{LN} is less than that for C_L. For the example V-n diagram calculations, it is assumed that $C_{LN} = 1.0$.

From equation 6.3.2, the −1g stall speed is 64.499 knots.

Equations 6.3.1 and 6.3.2 were used to graphically construct the V-n diagram shown in Figure 6.3.1. This diagram includes the safe load factor limits and the safe load factor the airplane can sustain at various speeds.

The limit load factors selected for the example V-n diagram are $n = +3.8$ and $n = -1.5$.

Thus, the airspeed at the maximum positive limit load factor of +3.8 is

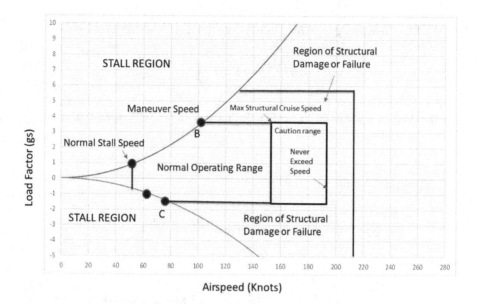

FIGURE 6.3.1 Velocity-load factor diagram.

$$V_R = \sqrt{\frac{2nW}{C_L \rho A_r}} = \sqrt{\frac{(2)(3.8)(2450)}{(1.42)(.0023767)(174)}} = 178.0679 \, \text{ft} / \text{sec} =$$

$$= 54.3 \, \text{m} / \text{sec} = 105.50 \, \text{knots}$$

and the airspeed at the maximum negative limit load factor of -1.5 is

$$V_C = \sqrt{\frac{2nW}{C_{LN} \rho A_r}} = \sqrt{\frac{(2)(1.5)(2450)}{(1.0)(.0023767)(174)}} = 133.3168 \, \text{ft/sec} =$$

$$= 40.6 \, \text{m} / \text{sec} = 78.987 \, \text{knots}$$

As shown in Figure 6.3.1, the intersection of the positive limit load factor and the line of maximum lift (point *B*) defines the maximum airspeed that allows full maneuverability. This point is called the maneuver speed. At lower speeds the aircraft structure cannot be overstressed as it will stall before reaching the limit load factor. At higher speeds, it may result in structural damage.

The intersection of the negative limit load factor and the line of maximum negative lift capability (point *C*) defines the maximum airspeed that allows full maneuverability for negative lift. Airspeeds greater than point *C* result in sufficient negative lift to damage the structure.

6.4 THE INSIDE LOOP

The loop is basically flight that follows a vertical circle path. We will begin the discussion of the loop by describing an inside loop where the maneuver is started by pitching upward. Starting from level flight, the inside loop is initiated by pulling back on the yoke or stick resulting in the nose coming up until all you see is sky. Then keep pulling until all you see is ground. Then continue pulling to return to level flight. It will be noted that the loop will not be a circle, but rather a teardrop shape if precise control of G forces (the load factor×the airplane's weight) as a function of θ (the angle between the wing chord and the horizon) is not maintained. The geometry of the loop and the inertial reference system forces and accelerations acting are shown in Figure 6.4.1. Note that a free body force diagram shows only the actual forces which are T, D, $mg = W$, and the lift force L.

If the thrust force (T) and the drag force (D) are equal, then there are no non-conservative forces acting on the airplane in the tangential direction.

The equations of motion are developed for any height $h = r(1 - \cos\theta)$ above the bottom of the loop using Newton's second law of motion.

The radial direction equation of motion follows, where L is the airplane lift force.

$$ma_r = m\frac{v^2}{r} = L - mg\cos\theta$$

$$L = m\frac{v^2}{r} + mg\cos\theta = m(\frac{v^2}{r} + g\cos\theta)$$

$$\text{radial direction load factor} = \frac{v^2}{gr} + \cos\theta$$

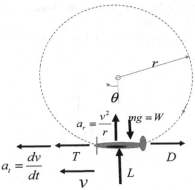

FIGURE 6.4.1 Inside loop path, forces, and accelerations (airplane loop initiation point shown).

The tangential direction equation of motion follows:

$$ma_t = m\frac{dv}{dt} = T - D - mg\sin\theta$$

if $T - D = 0$,

$$a_t = \frac{dv}{dt} = -g\sin\theta$$

tangential direction load factor $= -\sin\theta$

The total load factor is:

$$\sqrt{(\sin\theta)^2 + (\frac{v^2}{gr} + \cos\theta)^2} \qquad (6.4.1)$$

The load factor is the number of *gs* experienced by the airplane (object).

The airplane and its payload performing the loop can be weightless (zero load factor) in a zone of the maneuver where the lift force L is zero and thus the airplane and its payload are in free fall. The only force acting is thus the force of gravity.

Referring to Figure 6.4.2, the forces acting on the pilot in the airplane frame of reference are shown.

The forces acting on the pilot in the moving frame of reference (the airplane) shown in Figure 6.4.2 at the bottom and top positions of the loop reflect the inertia forces that result from the radial and tangential accelerations of the airplane. The free body diagram forces are the centrifugal force, $mg = W$, the reaction force F, and the tangential inertia force $m\frac{dv}{dt}$.

As shown, the airplane enters the loop at the bottom and this is where the loop starts as indicated by the subscript S.

From equilibrium, in the radial direction at any height $h = r(1 - \cos\theta)$ above the bottom of the loop:

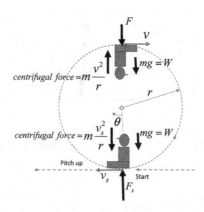

FIGURE 6.4.2 Inside loop forces acting on the pilot in the airplane frame of reference.

$$F - m\frac{v^2}{r} - mg\cos\theta = 0$$

where $m\dfrac{v^2}{r}$ is the radial direction inertia (centrifugal) force and F is the radial direction force exerted by the seat on the pilot.

In the tangential direction:

$$-m\frac{dv}{dt} - mg\sin\theta = 0$$

Note that the force equations are identical to those obtained using an inertial frame of reference where F in the airplane frame of reference replaces L in the inertial frame of reference.

The velocity at any height h above the bottom of the loop will be determined from the conservation of energy.

$$h_s mg + \frac{1}{2}mv_s^2 = \frac{1}{2}mv_f^2 + h_f mg$$

$$\frac{1}{2}mv_s^2 = \frac{1}{2}mv_f^2 + h_f mg - h_s mg = \frac{1}{2}mv_f^2 + mg(h_f - h_s) = \frac{1}{2}mv^2 + mgh$$

$$\frac{1}{2}mv_s^2 = \frac{1}{2}mv^2 + hmg$$

$$(6.4.2)$$

$$v = \sqrt{v_s^2 - 2gh} = \sqrt{v_s^2 - 2gr(1 - \cos\theta)}$$

This equation provides the velocity v in terms of the starting velocity v_S at the bottom of the loop.

To determine the loop entry velocity that results in zero gs at the top of the loop (the pilot and airplane are weightless), set the load factor to zero in equation 6.4.1.

$$\sqrt{(\sin 180)^2 + \left(\frac{v^2}{gr} + \cos 180\right)^2} = \frac{v^2}{gr} - 1 = 0$$

Thus, $v = \sqrt{rg}$ = velocity at the top of the loop.

From equation 6.4.2,

$$v = \sqrt{rg} = \sqrt{v_s^2 - 2gr(1 - \cos 180)} = \sqrt{v_s^2 - 4gr}$$

$$v_s = \sqrt{5rg} = \text{entry velocity}$$

and from equation 6.4.1,

$$\text{number of } gs \text{ at the bottom} = \left(\frac{v_s^2}{rg} + \cos 0\right) = \left(\frac{5rg}{rg} + 1\right) = 6$$

This means that to fly a circular loop without negative gs at the top, you need to pull at least 6*gs* at the bottom (loop entry point). This *g*-loading exceeds the structure failure load for small general aviation aircraft. This *g*-loading acts in the head to foot direction. It will be noted that this model is very conservative, as in reality, airplanes need much less than 6*gs* to do a loop. Typically, about +3.5*gs* is required to start the loop. The main reason for this is that the airplane will not slow down as fast going up the loop as the simple model indicates because of the non-conservative thrust force developed by the engine.

Note that for positive *gs*, blackout can occur in the range of 4–5 *gs* and unconsciousness (no blood in the brain) can occur in the range of 5–6 *gs*. These effects are variable from individual to individual, and some pilots can easily tolerate 6–7 positive *gs*, while others may lose consciousness at 3 or less *gs*.

Figure 6.4.3 shows plots of the inside loop tangential velocity and *g* level as a function of pitch angle for a loop entry velocity of 120 knots and zero *gs* at the top ($\theta = 180°$). These plots were generated by selecting the entry velocity equal to 120 knots $= 202.537$ feet/sec and using $v_s = \sqrt{5rg}$ to calculate $r = 255.015\,\text{ft} = 77.728\,\text{m}$.

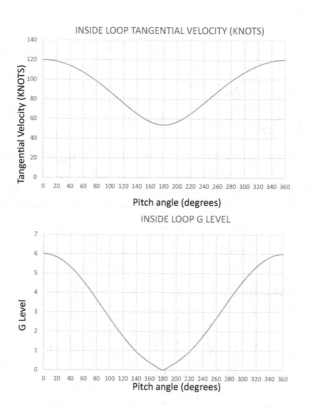

FIGURE 6.4.3 Inside loop tangential velocity and G level for an entry velocity of 120 knots and zero gs at top.

Note the cusp in the g level plot at $180°$ caused by the inclusion of the tangential acceleration in the load factor equation.

6.5 OUTSIDE LOOP

The outside loop maneuver can be initiated by pitching downward (pitch over) as shown in Figure 6.4.1. Note that this is the most hazardous way of performing the maneuver because the negative g levels and airspeed are increasing as the bottom of the loop is approached. This maneuver was previously thought to be a fatal maneuver and was performed by the American General and aviation pioneer James Doolittle in 1927 [13]. He made early coast-to-coast flights, earned a doctorate from M.I.T. in aeronautics, won many flying races, and most significantly, helped develop instrument flying. He was the commander of the Doolittle raid on Japan in World War II. The geometry of the loop and the inertial reference system forces and accelerations are shown in Figure 6.5.1. Note that a free body diagram shows only the actual forces which are T, D, $mg = W$, and L.

If the thrust force (T) and the drag force (D) are equal, then there are no non-conservative forces acting on the airplane in the tangential direction.

The equations of motion are developed for any distance $h = r(1 - \cos\theta)$ below the top of the loop using Newton's second law of motion.

The radial direction equation of motion follows, where L is the airplane lift force.

$$ma_r = m\frac{v^2}{r} = -L + mg\cos\theta$$

$$L = -m\frac{v^2}{r} + mg\cos\theta = m\left(-\frac{v^2}{r} + g\cos\theta\right)$$

$$\text{radial direction load factor} = -\frac{v^2}{gr} + \cos\theta$$

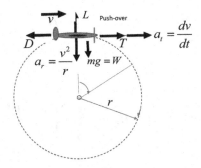

FIGURE 6.5.1 Outside loop path, forces, and accelerations (airplane loop initiation point shown).

The tangential direction equation of motion follows:

$$ma_t = m\frac{dv}{dt} = T - D - mg\sin\theta$$

if $T - D = 0$,

$$a_t = \frac{dv}{dt} = -g\sin\theta$$

tangential direction load factor $= -\sin\theta$

The total load factor is:

$$\sqrt{(\sin\theta)^2 + (-\frac{v^2}{gr} + \cos\theta)^2} \qquad (6.5.1)$$

The load factor is the number of gs experienced by the airplane (object).

The airplane and its payload performing the loop can be weightless (zero load factor) in a zone of the maneuver where the lift force L is zero and thus the airplane and its payload are in free fall. The only force acting is thus the force of gravity.

Referring to Figure 6.5.2, the forces acting on the pilot are shown in the airplane frame of reference.

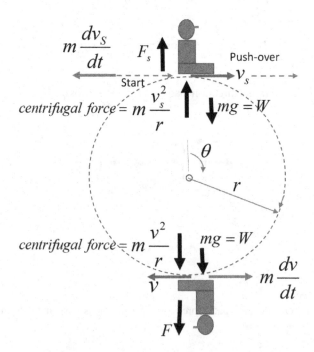

FIGURE 6.5.2 Outside loop forces acting on the pilot in the airplane frame of reference.

The free body diagram forces are the centrifugal force, $mg = W$, the reaction force F, and the tangential inertia force $m\dfrac{dv}{dt}$.

As shown, the airplane enters the loop at the top and this is where the loop starts as indicated by the subscript S.

From equilibrium, in the radial direction,

$$F + m\frac{v^2}{r} - mg\cos\theta = 0$$

where $m\dfrac{v^2}{r}$ is the radial direction inertia (centrifugal) force and F is the radial direction force exerted by the seat on the pilot and in the tangential direction:

$$-m\frac{dv}{dt} - mg\sin\theta = 0$$

Note that the force equations are identical to those obtained using an inertial frame of reference where F in the airplane frame of reference replaces L in the inertial frame of reference.

The velocity at any height $h = r(1 - \cos\theta)$ below the top of the loop is determined from the conservation of energy.

$$h_s mg + \frac{1}{2}mv_s^2 = \frac{1}{2}mv_f^2 + h_f mg$$

$$\frac{1}{2}mv_s^2 = \frac{1}{2}mv_f^2 + h_f mg - h_s mg = \frac{1}{2}mv_f^2 + mg(h_f - h_s) = \frac{1}{2}mv^2 - mgh$$

$$\frac{1}{2}mv_s^2 = \frac{1}{2}mv^2 - hmg$$

Thus,

$$v = \sqrt{v_s^2 + 2gh} = \sqrt{v_s^2 + 2gr(1 - \cos\theta)} \qquad (6.5.2)$$

To determine the loop entry velocity that results in zero gs at the top of the loop (the pilot and the airplane are weightless), set the load factor to zero in equation 6.5.1.

$$0 = \left(-\frac{v_s^2}{gr} + \cos 0\right) = \left(-\frac{v_s^2}{gr} + 1\right)$$

Then the entry velocity at the top of the loop is

$$v_s = \sqrt{rg}$$

The velocity at the bottom of the loop is

$$v = \sqrt{rg + 2gr(1 - \cos 180)} = \sqrt{rg + 4gr} = \sqrt{5gr}$$

The number of gs the bottom of the loop is

$$\left(-\frac{5rg}{rg} + \cos 180\right) = (-5 - 1) = -6$$

Figure 6.5.3 shows plots of the outside loop tangential velocity and g level as a function of pitch angle for a loop entry velocity of 120 knots and zero gs (weightless condition) at the top ($\theta = 0°$). These plots were generated by selecting the entry velocity equal to 120 knots = 202.537 ft/sec and using

$$v_s = \sqrt{rg} \text{ to calculate } r = 1,275.07 \text{ ft} = 388.641 \text{ m}.$$

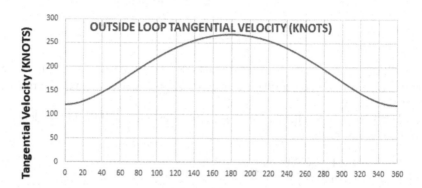

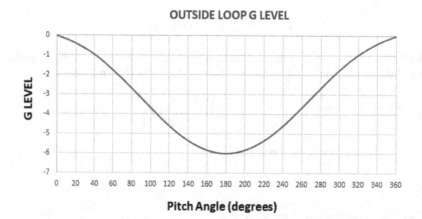

FIGURE 6.5.3 Outside loop tangential velocity and G Level for an entry velocity of 120 knots and zero gs at pitch over.

The resulting $-6gs$ at the bottom of the loop acts in the foot to head direction. This g level is beyond the structure failure load for a small general aviation aircraft.

Note that the slope of the g level is not zero at zero degrees because of the inclusion of the tangential acceleration in the load factor equation.

Note that negative gs are the least tolerated by humans and human tolerance to negative gs may be as low as $3gs$ for 5 seconds. However, the effects on negative gs are variable from an individual to individual.

Vision blurring can begin in the range of 1–2.5 negative gs, extreme discomfort can occur in the range of 2.5–3 negative gs, and incapacitation can occur for more than 3 negative gs.

It will be noted that many air show performers start the outside loop maneuver by simply rolling inverted to straight and level flight and then pitching downward (pushing the stick forward) and thus initiating the maneuver at the bottom of the circle shown in Figure 6.5.1. Typically, the maneuver can be started with as little as $-4gs$.

Also note that this approach avoids the inherent danger of increasing negative gs that result from starting the maneuver by pitching over at the top of the loop. This method has the advantage of decreasing air speed as the air plane climbs up the loop and this results in lower negative gs over the complete profile.

6.6   AILERON ROLL

The aileron roll is a flight maneuver in which the airplane is rotated about its longitudinal axis through a full 360° by means of the ailerons. The longitudinal axis vector of the airplane can be in any direction including horizontal, vertical, 45°, etc. If the altitude of the airplane is not altered, then the maneuver is referred to as a slow roll. Considering a constant altitude aileron roll and using the aircraft frame of reference, as the airplane is rolled through the $1g$ field, the $1g$ vector acts in the head to foot direction of the pilot (positive $1g$) at the initiation of the roll and at 180° and the $1g$ vector acts in the foot to head direction (negative $1g$). Figure 6.6.1 shows the G force (mg) vector relative to the airplane at zero and at 180° of roll angle. At these roll angles, the G force acts in the direction perpendicular to the wings.

At all intermediate roll angles, the $1g$ vector has components in directions perpendicular and parallel to the wing axis which are equal respectively to $mg\cos\theta$ (pilot's head to foot or foot to head direction) and $mg\sin\theta$ (pilot's side direction).

To review: at the start of the roll, the pilot experiences $1g$ head to foot and 180° later, the pilot experiences $-1g$ foot to head. At intermediate roll angles, the pilot experiences both foot to head or head to foot fractional gs as well as $1g$ or fractional side gs.

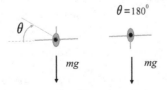

FIGURE 6.6.1   Aileron roll G force vector relative to the airplane at zero and 180° of roll angle.

6.7   THE BARREL ROLL

The barrel roll is a combination of a loop and roll executed simultaneously. This maneuver is referred to as the barrel roll because the airplane's flight path is not in a straight line, as in a pure aileron roll or slow roll, but along a helical path, as if it were spiraling along the surface of an invisible cylinder or barrel. Roll rate, airspeed, and pitch attitude are constantly changing during the maneuver.

   The centrifugal force generated by the curved flight path can keep the g level positive (acting in the head to foot direction) throughout the maneuver if the roll rate is high enough such that it's possible that a glass of water on the aircraft's glare shield should remain in position, and the level of the water should remain parallel to the wing.

   Figure 6.7.1 shows barrel roll projection views along the airplane's longitudinal (air speed direction) or forward velocity component (velocity into the page) showing the airplane tangential position direction component.

   In Figure 6.7.1, the three airplane location depictions are shown at three different times. The total motion of the airplane follows a helical flight path.

   In cylindrical coordinates, the displacements of the airplane executing the barrel roll are

$$r, \theta, ut$$

Where u is the airplane longitudinal velocity component (air speed).

   Radial acceleration =

$$a_r = \frac{v^2}{r} = \frac{r^2 \omega^2}{r} = r\omega^2 = r\dot{\theta}^2$$

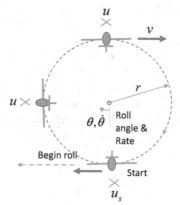

v = airplane tangential velocity component

u = airplane longitudinal velocity component where X is the tail of the arrow representing the longitudinal velocity vector component

FIGURE 6.7.1   Barrel roll airplane orientations and velocity components.

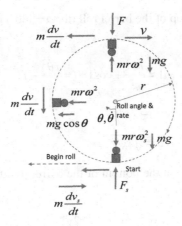

FIGURE 6.7.2 Barrel Roll radial and tangential forces in the airplane frame of reference.

Referring to Figure 6.7.2, in the airplane frame of reference, force equilibrium in the radial direction at any height from the bottom of the barrel roll gives:

$$F - mr\omega^2 - mg\cos\theta = 0$$

$$F - mr\dot{\theta}^2 - mg\cos\theta = 0$$

$$F = mr\dot{\theta}^2 + mg\cos\theta = m(r\dot{\theta}^2 + g\cos\theta)$$

Radial direction load factor =

$$= \frac{r\dot{\theta}^2}{g} + \cos\theta$$

and force equilibrium in the tangential direction is:

$$-m\frac{dv}{dt} - mg\sin\theta = 0$$

Tangential acceleration =

$$a_t = \frac{dv}{dt} = -g\sin\theta$$

Tangential load factor =

$$-\sin\theta$$

The total load factor is:

$$\sqrt{(\sin\theta)^2 + (\frac{r\dot{\theta}^2}{g} + \cos\theta)^2} \qquad (6.7.1)$$

If the load factor at the top of the barrel roll (theta = 180°) is selected to be 1.0 then from equation 6.7.1:

$$1 = \frac{r\dot\theta^2}{g} + \cos 180 = \frac{r\dot\theta^2}{g} - 1$$

$$\dot\theta^2 = 2g/r$$

$$\dot\theta = \sqrt{2g/r}$$

If this is also the roll rate at the bottom of the barrel roll (theta = 0°), then the load factor is:

$$\frac{r\dot\theta^2}{g} + \cos 0 = \frac{r2g}{gr} + 1 = 3$$

Thus, if the roll rates are equal at the top and bottom of the barrel roll, although in fact the roll rate as discussed below actually changes over the maneuver, then a load factor of +1.0 at the top requires a load factor of +3.0 at the bottom.

Note that the minimum positive load factor occurs at the top and the maximum positive load factor occurs at the bottom.

Thus in a proper barrel roll (circular tangential flight path), a uniform load factor, say 1.0, is impossible to obtain.

Therefore, while a barrel roll can be a positive g maneuver if the roll rate is high enough, it cannot be a 1g maneuver.

As a point of reference in history, Alvin "Tex" Johnston, the Boeing test pilot, performed a barrel roll of Boeing's new Dash 80, the prototype of the 707 commercial airliner on August 7, 1955 (Ref. [11]).

It was stated that this was a +1g maneuver. However, as shown herein, while the barrel role can produce positive gs, it cannot be a 1g maneuver.

Also note that the roll rate at the bottom is actually greater than the roll rate at the top of the barrel roll. Thus it is concluded that to obtain a load factor of +1.0 at the top, a load factor greater than +3.0 is required at the bottom.

For example, selecting the barrel roll radius equal to 200 ft = 60.96 m, the roll rate for 1g at the top is:

$$\dot\theta = \sqrt{2g/r} = \sqrt{\frac{2 \times 32.2}{200}} = .567\,\text{rad/sec}$$

and .567 rad/sec × 57.295 deg/rad = 32.5 deg/sec

Selecting a reasonable roll rate at the bottom of 40 deg/sec and assuming that it lowers to 32.5 deg/sec at the top result in a load factor of 4.0 gs at the bottom.

The translational and angular velocities at any height h above the bottom of the barrel roll can be quantified using the conservation of energy:

$$\frac{1}{2}mr^2\dot\theta_s^2 + \frac{1}{2}mu_s^2 + h_s mg = \frac{1}{2}mr^2\dot\theta_f^2 + \frac{1}{2}mu_f^2 + h_f mg$$

$$\frac{1}{2}mr^2\dot\theta_s^2 + \frac{1}{2}mu_s^2 = \frac{1}{2}mr^2\dot\theta^2 + \frac{1}{2}mu^2 + hmg$$

Starting velocities have the subscript S and velocities with the subscript f for a given theta are shown without the subscript and the height above the bottom of the barrel roll is $h = r(1 - \cos\theta)$.

$$(\dot\theta^2 + \frac{u^2}{r^2}) = (\dot\theta_s^2 + \frac{u_s^2}{r^2}) - \frac{2g}{r}(1 - \cos\theta)$$

$$u = \sqrt{(\dot\theta_s^2 - \dot\theta^2)r^2 + u_s^2 - 2gr(1 - \cos\theta)} \tag{6.7.2}$$

Selecting an entry air speed to the barrel roll at theta$=0°$ of u_s equal to 120 knots and using the bottom and top roll rates of 40 and 32.5 deg/sec, respectively, then from equation 6.7.3, the velocity (air speed) at the top of the barrel roll (theta$=180°$) is equal to 87.7 knots. Thus the airspeed reduces from 120 knots at the bottom of the barrel roll to 87.7 knots at the top of the barrel roll.

6.8 PRECESSION DUE TO PITCH ANGULAR VELOCITY

One of the turning or yaw forces (actually, moments) acting on an aircraft is due to pitch angular velocity of the aircraft's propulsion system rotor longitudinal axis. Note that the combined system of the propeller and engine rotor or only the rotor of a jet engine is in essence a gyroscope rotor and a change in the yaw direction of the rotor's spin axis angular velocity vector direction such as due to aircraft pitching is called precession.

As a result of gyroscopic action, pitching of the rotor's spin axis results in a vertical axis or yaw direction moment vector.

Referring to Figure 6.8.1, the gyroscopic moments acting on the aircraft engine rotor in the vertical and horizontal planes relative to an inertial reference frame are given by the following equations:

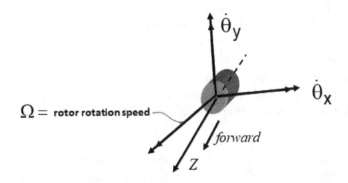

FIGURE 6.8.1 Aircraft engine spinning rotor and ground based angular velocity directions.

$$-\Omega I_p \dot{\theta}_Y = M_X \qquad (6.8.1)$$

$$\Omega I_p \dot{\theta}_{YX} = M_Y \qquad (6.8.2)$$

Note that equal and opposite moments are acting on the engine supporting static structures (from Newton's third law) and thus the gyroscopic moments acting on the aircraft are equal in sign and magnitude to those provided by equations 6.8.1 and 6.8.2.

When writing Newton's second law equations of motion, these forces are added to the stiffness and damping forces to permit the complete solution for dynamic displacements and reaction forces resulting from applied forces.

Equations 6.8.1 and 6.8.2 express the cross coupling that exists between the angular velocity and moment vectors in orthogonal directions.

In Figure 6.8.1, the forward direction is toward the left.

Ω rotor spin speed (clockwise rotation direction, aft looking forward)
$\dot{\theta}_X$ the angular velocity about the X axis (pitch axis)
$\dot{\theta}_Y$ the angular velocity about the Y axis (yaw axis)
M_X the moment about the X axis (pitch axis)
M_Y the moment about the Y axis (yaw axis)
I_X and I_Y are the rotor diametrical mass moments of inertia
I_P is the rotor polar mass moment of inertia

Note that M_X and M_Y are moment vectors pointed in the horizontal and vertical directions, respectively.

From equation 6.8.2, for a pitch down angular velocity (positive $\dot{\theta}_{YX}$), the yaw moment M_Y acts to turn the aircraft in the counter clockwise direction (looking down).

This is the case for a tailwheel aircraft, where when the tail is raised on the take-off roll (hence the nose is lowered), there is a moment tending to turn the aircraft to the left.

Note from equation 6.8.1 that the angular velocity about the y axis (yaw axis) causes a moment about the x axis (pitch axis).

Also note that counter rotating propellers were used in the Wright Brothers first airplane such that yawing of the airplane was canceled for pitching of the propeller's axes and thus that of the airplane longitudinal axis.

PROBLEMS

1. What unbalanced force causes an airplane to turn?
2. What is the airplane primary turn control?
3. What are the forces acting on an airplane in both coordinated and uncoordinated level turns?
4. What is an airplane slip or skid, what causes this behavior in uncoordinated turns, and what can be done to avoid these phenomena?
5. Describe the effect of bank angle on the stall speed.

6. Describe the velocity-load factor diagram, its purpose, and how it is generated.
7. Define maneuver speed.
8. When can the condition of free fall and thus weightlessness occur in a loop and what is the governing equation defining the airplane's velocity for this condition?
9. Why does the pilot experience negative gs when an aileron roll is performed?
10. Describe why a uniform load factor of +1 cannot be obtained in a barrel roll.
11. Explain why counter rotating propellers were used in the Wright Brothers airplanes.

7 Propulsion

Most general aviation aircraft uses reciprocating engines based on performance and cost considerations. Regional transport aircraft typically uses turboprop engines based on relatively shorter route requirements and higher fuel efficiency. Larger transport aircraft uses jet engines to meet higher power and longer route requirements.

All aircraft propulsion systems, propeller-reciprocating engine, propeller-gas turbine (turboprop) engine, or jet engine, have in common the way they produce thrust. Regardless of the type of propulsion system, the development of thrust is governed by Newton's second law of motion as follows:

$$F = ma \text{ or } F = \frac{d\ mv}{dt} = m\frac{dv}{dt} \text{ if mass is constant.}$$

where F is the working fluid accelerating force (an equal and opposite force acts on the aircraft) and mv is the momentum of the working fluid.

The thrust force results from the acceleration provided to the mass of the working fluid through the action of pressure and frictional forces. The magnitude of the thrust force is dependent on the rate of change of the momentum of the working fluid.

7.1 PROPELLER PROPULSION

Referring to Figure 7.1.1, and considering the conservation of mass at the entry and exit stations of the control volume (shown encased in dashed lines):

$$\rho_0 A_0 V_0 = \rho_e A_e V_e = \dot{m} = \text{mass flow rate.}$$

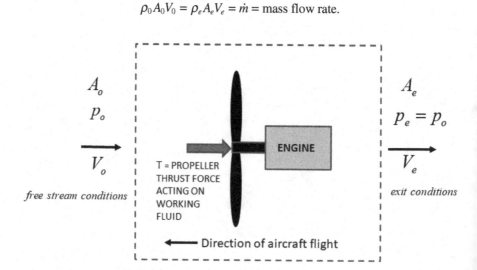

A_o

p_o

V_o

free stream conditions

T = PROPELLER THRUST FORCE ACTING ON WORKING FLUID

ENGINE

A_e

$p_e = p_o$

V_e

exit conditions

← Direction of aircraft flight

FIGURE 7.1.1 Propeller and engine control volume.

DOI: 10.1201/9781003390916-7

The momentum rate of change at the entry and exit stations is

$$\rho_0 A_0 V_0 V_0 = \dot{m} V_0 \text{ and } \rho_e A_e V_e V_e = \dot{m} V_e, \text{respectively.}$$

Therefore, the momentum rate of change across the control volume is

$$\dot{m} V_e - \dot{m} V_0 = \dot{m}(V_e - V_0).$$

From Newton's second law of motion assuming that the exit pressure is equal to the free stream pressure,

$$T = \dot{m}(V_e - V_0) \tag{7.1.1}$$

where T is the net thrust force acting on the working fluid. Obviously, an equal and opposite force acts on the propeller or on a jet engine and thus the aircraft (Newton's third law of motion). Note that V_0 is the velocity of the air entering the plane of the propeller and is equal and opposite to the velocity of the aircraft (the true airspeed of the aircraft) and results in a loss of thrust.

The speed of the exit air relative to the engine or aircraft is V_e. The term $\dot{m} V_0$ is known as the ram drag and $\dot{m} V_e$ is the static thrust.

The air entering the propeller is accelerated resulting in an increase of the momentum rate of change of the air and thus a thrust or accelerating force acts on the air as discussed above. Presented herein are the details of how the propeller is used to accomplish the momentum rate of change. In this development, the thrust force T will be defined as that acting on the propeller. The propeller blade is actually a strongly twisted wing as discussed by Von Mises, Richard in Reference [15].

The cross sections of the propeller blade are essentially of the same shape as those of a wing, with a rounded leading edge and a sharp trailing edge. However, unlike a wing where the chord lines of the airfoil sections are essentially at the same orientation, a propeller is twisted such that the angle of the chord lines of the cross sections varies along the length of the propeller blade. The angle β that the propeller cross sections chords form with a plane perpendicular to the axis of the propeller is much greater for the sections near the hub than for those near the blade tip. The large blade twist is necessary to ensure that a favorable angle of attack α is obtained for each blade cross section.

As shown in Figure 7.1.2, the propeller axis of rotation coincides with the direction of flight and each blade cross section has a velocity component in the direction of the propeller axis V_0 and a rotational velocity component V_t parallel to the plane perpendicular to the axis of the propeller. The velocity component V_0 in the direction of the propeller axis obviously has the same value for all cross sections, while the perpendicular velocity component V_t for each cross section has a value that is proportional to the radial distance from the propeller axis. The velocity component V_t is equal to the rotational speed ω (angular velocity) of the propeller times the radial direction r from the propeller axis.

The angle ϕ between the plane of rotation and the resultant velocity V is determined from

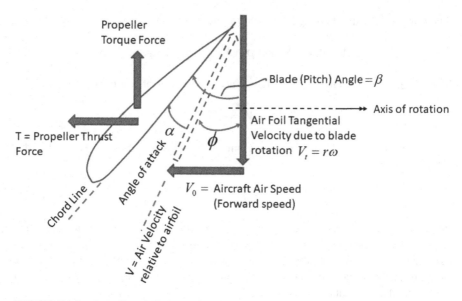

FIGURE 7.1.2   Propeller airflow and force vectors.

$$\tan\phi = \frac{V_0}{r\omega} \tag{7.1.2}$$

Thus, the angle ϕ decreases with increasing r and due to the blade twist, the blade angle β is reduced for increasing r so that the angle of attack α is not increased. The change in blade angle β or the "twisting" along the length of the propeller permits the propeller to operate with a relatively uniform angle of attack along its length when in cruising flight.

When specifying the pitch for a fixed pitch propeller, the propeller manufacturer usually selects a pitch that will allow efficient operation at the expected airplane cruising speed and the corresponding engine RPM. Unfortunately, every fixed pitch propeller must be a compromise because it can be efficient only at a given combination of airspeed and RPM.

At the propeller blade tip, ϕ_1 is defined as:

$$\tan\phi_1 = \frac{2V_0}{d\omega}$$

where d is the diameter of the propeller

Then

$$\tan\phi = \frac{V_0}{r\omega} = \frac{d}{r}\frac{V_0}{\omega d} = \frac{d}{2r}\tan\phi_1$$

The angle of attack α between the airfoil chord line and the relative velocity V for a given value of r is given by

$$\alpha = \beta - \phi = \beta - \tan^{-1}\left(\frac{d}{2r}\tan\phi_1\right) \tag{7.1.3}$$

For the propeller airfoil, the lift force L and the drag force D are respectively perpendicular and parallel to the airfoil relative velocity V. Thus, from Figure 7.1.2, the propeller airfoil thrust and torque forces are as follows:

$$\text{Thrust Force} = T = L\cos\phi - D\sin\phi$$

$$\text{Torque Force} = L\sin\phi + D\cos\phi \tag{7.1.4}$$

In order to provide the blade angle β to obtain the most efficient angle of attack over different airspeed and engine power conditions, variable pitch propellers are used whereby the blade can be rotated around its long axis to change the blade pitch. As the blade angle is increased, the propeller produces more thrust force and an increase in engine torque is required. As the blade angle is decreased, less thrust force results and less engine torque is required.

A type of variable pitch propeller is the constant speed propeller which allows the pilot to set the desired engine speed and the propeller blade pitch is automatically controlled so that the engine speed remains constant irrespective of the amount of engine torque being produced. The main advantage of a constant speed propeller is that it converts a high percentage of the engine's power into thrust over a high range of engine rotational speeds and airspeed combinations.

Note that for engine failure, the propeller control mechanism allows the propeller to be feathered to reduce propeller drag (reversed thrust). The control mechanism permits the pitch angle to be changed to 90°. Therefore, the propeller chord is rotated so that it is parallel to the direction of flight and thus the propeller stops rotating and minimum windmilling, if any, occurs.

7.2 JET ENGINE PROPULSION

Figure 7.2.1 shows a simplified sketch of a turbojet engine. Note that both solid and gaseous boundaries of the control volume are shown. For the solid boundaries, airflow cannot pass, and for the gaseous boundaries, airflow can pass. In Figure 7.2.1, stations 0 and e represent open boundaries through which airflow can pass and the only forces that can act on them are pressure forces. The other boundaries are solid walls or the jet engine casing and jet engine blades, rotors, frames, and internal structures on which both pressure and friction forces act. These forces produce the engine – working fluid force, that is, the engine thrust force T acting on the working fluid. Obviously, an equal and opposite force acts on the jet engine and thus the aircraft (Newton's third law of motion). Note that air is the working fluid and not fuel. Refer to Reference [7] for a further discussion of the control volume model.

Newton's second law of motion provides for the control volume:

$$T - p_e A_e + p_0 A_0 - p_0(A_0 - A_e) = \dot{m}(V_e - V_0) \text{ or}$$

$$T - (p_e - p_0)A_e = \dot{m}(V_e - V_0) \text{ or} \tag{7.2.1}$$

$$T = \dot{m}(V_e - V_0) + (p_e - p_0)A_e$$

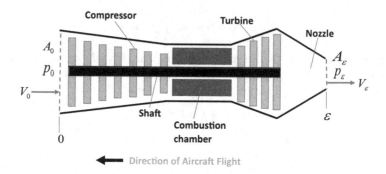

FIGURE 7.2.1 Turbojet engine.

where p_e is the exit static pressure, p_0 is the air ambient static pressure, and $T - (p_e - p_0)A_e$ is the sum of the forces that accelerate the working fluid.

The following is a numerical example of the thrust calculation for an operating aircraft turbojet engine.

Barometric pressure $= p_0 = 29.34$ inches of mercury $= 14.4 \dfrac{\text{lbs}}{\text{in}^2} = 2{,}073.6 \dfrac{\text{lbs}}{\text{ft}^2}$

Aircraft true airspeed $= V_0 = 269$ knots $= 460$ ft/sec
Compressor weight flow rate $= \dot{w} = 96$ lbs/sec
Exhaust nozzle area $= A_e = 2 \text{ ft}^2$
Exhaust nozzle static pressure $= p_e = p_0$, assuming that the exhaust jet expands
to ambient pressure
Exhaust gas velocity $= V_e = 1{,}460$ ft/sec. From equation 7.2.1:

$$T = \dot{m}(V_e - V_0) + (p_e - p_0)A_e = \frac{96}{32.2}(1{,}460 - 460) = 2{,}981.366 \ lbs$$

From the thrust equation 7.2.1, increased thrust can be obtained by increasing the exit velocity of the exhaust gas or by increasing the air mass flow rate. The latter choice is used in high bypass-ratio turbofan engines (Figure 7.2.2) to achieve increased thrust at high propulsive efficiency (the work done to propel the aircraft divided by the work done by the engine to accelerate the engine air mass).

The bypass ratio of a turbofan engine is the ratio between the air mass flow rate of the bypass stream and the air mass flow rate entering the core

$\dot{m}_f =$ bypass air mass flow rate
$\dot{m}_c =$ core air mass flow rate
$V_0 =$ aircraft velocity
$V_f =$ air velocity at fan discharge $=$ velocity at inlet to low-pressure compressor
(Booster)
$V_{fe} =$ air velocity of bypass stream
$V_c =$ core gas velocity at core exit
$P_{fe} =$ bypass air discharge pressure
$P_c =$ core gas discharge pressure

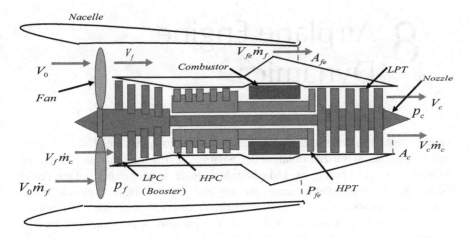

FIGURE 7.2.2   Turbofan engine.

LPC =Low-pressure compressor (Booster)
HPC = High-pressure compressor
LPT =Low-pressure turbine
HPT =High-pressure turbine
N_1 =Fan speed
N_2 =Core speed

From Newton's second law of motion:

$$T = \dot{m}_f(V_{fe} - V_0) + \dot{m}_c(V_c - V_f) + (P_{fe} - P_0)A_{fe} + (P_c - P_0)A_c$$

$$BPR = \frac{\dot{m}_f}{\dot{m}_c} = \text{Bypass ratio}$$

$$T = BPR[\dot{m}_c(V_{fe} - V_0)] + \dot{m}_c(V_c - V_f) + (P_{fe} - P_0)A_{fe} + (P_c - P_0)A_c \quad (7.2.2)$$

The addition of fuel mass flow rate has been neglected.

T includes the core and fan thrust forces acting on the working fluids and an equal and opposite force acts on the engine and thus the aircraft. Thus, both the fan and the core contribute to the total thrust and for a high bypass engine, the fan provides the major portion of the thrust because of its high mass flow rate.

PROBLEMS

1. Explain how Newton's second law is used to describe how thrust is produced for all aircraft propulsion systems.
2. Explain why variable pitch propellers are more efficient in producing thrust and define what propeller blade feathering is and how it is used to reduce propeller drag in the event of engine failure.
3. Explain the difference between turbojet and turbofan engines and why turbofan engines are more efficient.

8 Airplane Engine Dynamics

Another flight area dependent on Newtonian mechanics is the discipline of airplane engine dynamics. It is used to analyze the dynamics of structures that contain rotating components such as piston, jet, and gas turbine engines. As the speed of rotation of the propulsion system rotor increases, the amplitude of vibration often passes through a maximum at a speed that is called the critical speed. The amplitude of vibration is commonly excited by unbalance of the rotor. If the amplitude of vibration at the critical speed is excessive, then damage to the engine system can occur. Vibration behavior of the engine system due to unbalance is a major consideration in engine design. A simplified rotor model based on the Jeffcott analysis (Reference [8]) is presented that utilizes a simple, concentrated mass, and flexible rotor and includes the effect of external damping. This model demonstrates all of the important aspects of basic lateral synchronous (rotor spin or rotational speed equals the whirling speed or frequency of vibration) rotordynamic vibration including the phase shift of the rotor's center of mass as its critical speed is traversed. Jeffcott's original analysis is based on the rotor model shown in Figure 8.1. This model consists of a single concentrated mass located in the center of flexible shaft which is mounted on simple supports.

In the end view of the rotor model shown in Figure 8.1, the origin of the stationary (non-rotating) coordinate system XY is 0 and is located at the undeflected centerline of the rotor. The point C represents the deflected centerline of the rotor and the point M represents the location of the center of mass of the rotor. The distance between 0 and C represents the displacement of the rotor centerline and is identified as the value A. The distance between C and M is equal to ε and the angle between A and ε is the phase angle β.

The frequency of vibration (whirl speed) is ω rad/sec. The mass of the rotor is m and c and k are the external viscous damping and stiffness, respectively.

The forces $m\ddot{x}'$ and $m\ddot{y}'$ shown in the inertial coordinate system of Figure 8.2 are reversed effective or D'Alembert forces which are fictitious inertia forces that are placed on the left side of Newton's second law equations of motion (equation 8.1) to obtain the equations for an equivalent static system.

From Newton's second law and the free body diagram of Figure 8.2, the differential equations of motion are:

$$m\,\ddot{x}' + c\,\dot{x} + k\,x = 0$$

$$m\,\ddot{y}' + c\,\dot{y} + k\,y = 0$$

(8.1)

Shown in Figure 8.2 are the forces acting on the disk at an instant in time when the rotor is at an angle ωt from the X axis. The forces depicted in Figure 8.2 are shown in a

10.1201/9781003390916-8

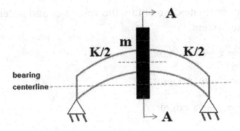

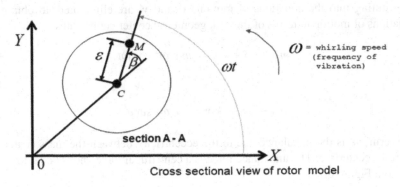

FIGURE 8.1 Simple rotordynamic model of a symmetric rotor with a center mounted disk (concentrated mass).

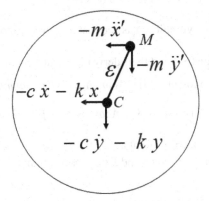

FIGURE 8.2 Disk free body diagram including fictitious inertia forces.

non-rotating (inertial) coordinate system. The geometric center coordinates of the disk C at the same instant in time are x and y. The coordinates of the disk center of mass M at the same instant in time are x' and y' related to x and y by the following equations.

$$x' = x + \varepsilon \cos \omega t$$

$$y' = y + \varepsilon \sin \omega t$$

Differentiating the center of mass coordinates, we obtain the velocities and accelerations in terms of the x, y coordinate system.

$$\dot{x}' = \dot{x} - \omega\,\varepsilon\,\sin\omega t$$

$$\ddot{x}' = \ddot{x} - \omega^2\varepsilon\cos\omega t$$

$$\dot{y}' = \dot{y} + \omega\varepsilon\cos\omega t$$

$$\ddot{y}' = \ddot{y} - \omega^2\varepsilon\sin\omega t$$

Substituting into the equations of motion, x' and y' are eliminated, to obtain the equations of motion in terms of the disk geometric center coordinates.

$$m\,\ddot{x} + c\,\dot{x} + k\,x = m\,\varepsilon\,\omega^2\,\cos\omega t$$

$$\text{(8.2)}$$

$$m\,\ddot{y} + c\,\dot{y} + k\,y = m\,\varepsilon\,\omega^2\,\sin\omega t$$

The term $m\varepsilon$ is the unbalance due to the eccentricity between the mass center and the rotor centerline. The unbalance force is a centrifugal force and is equal to $m\varepsilon\omega^2$.
From Figure 8.1,

$$x = A\cos(\omega t - \beta)$$

$$y = A\sin(\omega t - \beta) \qquad \text{(8.3)}$$

Where the amplitude of vibration A is the distance from 0 to C and β is the phase angle.

Substituting equations 8.3 and their derivatives into equations 8.2 gives

$$A(k - m\omega^2)\cos(\omega t - \beta) - Ac\omega\sin(\omega t - \beta) = m\varepsilon m\omega^2\cos\omega t$$

$$A(k - m\omega^2)\sin(\omega t - \beta) + Ac\omega\cos(\omega t - \beta) = m\varepsilon\omega^2\sin\omega t \qquad \text{(8.4)}$$

Figure 8.3 shows a graphical presentation of these equations that depicts the rotating vectors representing the forces and from observation, the values of A and β are obtained.

$$c^2A^2\omega^2 + A^2(k - m\omega^2)^2 = m^2\varepsilon^2\omega^4$$

$$A = \frac{m\varepsilon\omega^2}{\sqrt{(k - m\varepsilon\omega^2)^2 + c^2\omega^2}}$$

$$\beta = \arctan\left(\frac{c\omega}{k - m\omega^2}\right)$$

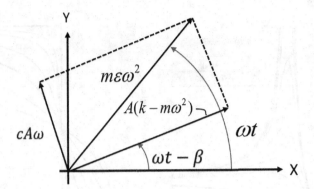

FIGURE 8.3 Graphical description of equations of motion whirling force vectors in an inertial frame.

In terms of non-dimensional parameters,

$$A = \frac{\varepsilon(\omega/\omega_n)^2}{\sqrt{(1-\omega^2/\omega_n^2)^2 + 4(\zeta\omega/\omega_n)^2}} \tag{8.5}$$

$$\beta = \arctan\left[\frac{2\zeta\omega/\omega_n}{1-\omega^2/\omega_n^2}\right] \tag{8.6}$$

$$\omega_n = \text{undamped natural frequency} = \sqrt{k/m}$$

$$\zeta = \text{critical damping ratio} = C/C_c$$

$$C_c = \text{critical damping} = 2\sqrt{k\,m} = 2\,m\,\omega_n$$

The damped critical speed of the rotor is defined as the speed at which the value A is a maximum. The equation for the damped critical speed is derived by differentiating equation 8.5 with respect to ω, setting it equal to zero, and solving for ω, resulting in

$$\omega_{cr} = \frac{2\,k}{\sqrt{4\,m\,k - 2\,c^2}} = \frac{\omega_n}{\sqrt{1-2\,\zeta^2}} = \text{damped critical speed}$$

It is apparent from this equation that the effect of damping is to increase the forced response critical speed.

Plots of the normalized response A/ε and the phase angle β as a function of frequency ratio and damping are presented in Figure 8.4.

As shown in Figure 8.5, when the rotor is spinning at a speed that is significantly greater than the undamped critical speed (super critical operation), then the rotor displacement A is equal to ε and opposite in direction. Note that for clarity in Figure 8.5, A is falsely shown to be larger than ε for this condition. When the rotor is spinning at a speed significantly below the undamped critical speed, the heavy side is outside of the rotor orbit. What is significant in Figures 8.4 and 8.5 is that at a spin speed equal

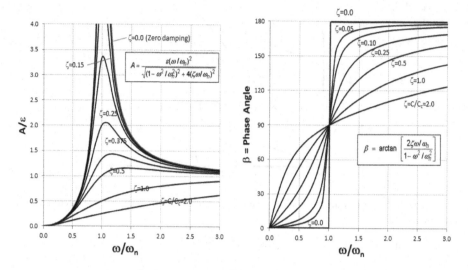

FIGURE 8.4 Non-dimensional amplitude and the phase angle for an unbalanced rotor as a function of rotor speed and damping in an inertial frame.

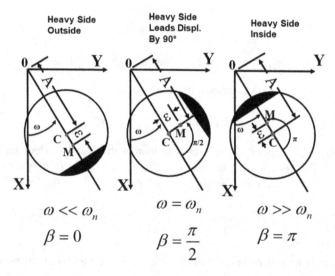

FIGURE 8.5 Phase angle between unbalance force and disk displacement as a function of rotor speed.

to the undamped critical speed, the unbalance force leads the rotor displacement vector by 90° as also shown in Figure 8.6, and for spin speeds significantly above the undamped critical speed, the mass center M moves to the undeflected center 0 of the rotor such that the heavy side is inside of the orbit and the geometric center C of the rotor is spinning and whirling about the mass center M. Thus, the centrifugal

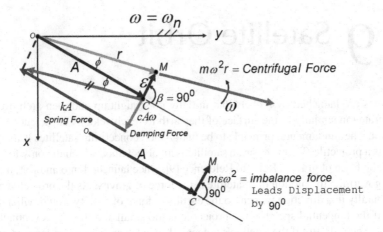

FIGURE 8.6 Vector diagram for shaft whirling at undamped critical speed.

force due to unbalance is equal to zero and no force is reacted by the rotor bearings. This is a condition of rotor isolation.

Note that in the above description, the rotor is spinning at an angular velocity equal to that of the orbit angular velocity and this is described as a synchronous orbit. An example of a synchronous orbit is that of the moon spinning about its axis and orbiting the earth with equal spin and orbital angular velocities resulting in the same side of the moon always facing the earth. The side of the moon not facing the earth is termed the far side of the moon.

PROBLEMS

1. Provide the definition of the critical speed for piston, jet, and gas turbine engines.
2. Show how the use of Newton's second law provides the dynamic equations of motion that address the effects of the rotating components in aero engines.
3. Discuss why engine lateral synchronous vibration response due to rotor unbalance is an important consideration in engine design.
4. Discuss the response behavior of the engine rotor when it is spinning at speeds below the critical speed, at the critical speed, and above the critical speed.

9 Satellite Orbit

Satellites are launched by rockets that initially go straight up and then arch over and accelerate tangentially to the surface of the earth to reach velocity sufficient to obtain an orbit. The fundamental principle to be understood concerning satellites is that a satellite is a projectile. That is to say, a satellite is an object upon which the only force acting is the force of gravity. Refer to Reference [9]. Once launched into an orbit, the only force governing the motion of a satellite is the force of gravity. As the projectile moves horizontally in a direction tangent to the earth, the force of gravity would pull it downward if the tangential speed of the projectile is too small and it would eventually fall toward the earth. But if the satellite's tangential speed is sufficient, the force of gravity acts as an unopposed turning force (acting perpendicular to the tangential velocity) such that sufficient centripetal acceleration is produced by the turning of the tangential velocity vector to keep the satellite moving in a circular path around the earth.

The edge of space begins at approximately an altitude of 100 km = 62 miles = 3.28083×10^5 feet. At this altitude, the atmosphere and thus the drag force acting on a satellite are basically non-existent. Note that the 100 km mark is termed the Karman line after Theodore von Karman.

The radius of the earth r_0 is 6,372 km = 3,959 miles = 2.09022×10^7 ft. For an altitude of 100 km, the distance to the center of the earth is:

$$r = r_0 + 3.28083 \times 10^5 = 2.09022 \times 10^7 + 3.28083 \times 10^5 = 2.123028 \times 10^7 \, \text{ft}$$

and the gravitational acceleration $g = 31.1206 \, \text{ft/sec}^2$ compared to $g = 32.2 \, \text{ft/sec}^2$ at the surface of the earth.

In an inertial frame (the earth), the accelerating centripetal gravitational force F_G causes the satellite to follow a curved path and results in centripetal acceleration $\frac{V^2}{r}$. Equating the gravitational force to the centripetal acceleration × mass (Newton's second law), the tangential velocity V required to maintain the orbit is obtained.

$$\frac{GMm}{r^2} = m\frac{V^2}{r}$$

$F_G = \dfrac{GMm}{r^2}$ is the Newton inverse square law of gravitation,

where F_G is the gravitational force acting on the satellite, G is the universal gravitational constant, M is the mass of the earth, m is the mass of the satellite, and r is the distance between the satellite and the center of the earth

$$F_G = \frac{GMm}{r^2} = \frac{\mu m}{r^2} = \frac{gr^2}{r^2}m = mg$$

DOI: 10.1201/9781003390916-9

where μ is the gravitational parameter of the earth.

Thus,

$$mg = m\frac{V^2}{r}$$

where mg is the gravitational force (weight or the accelerating force acting on the satellite).

Solving for the satellite velocity,

$$a = \frac{V^2}{r}\ \text{ft/sec}^2 = \text{centripetal acceleration}$$

$$V = (gr)^{1/2} \tag{9.1}$$

Also note that the acceleration due to gravity g is equal to the centripetal acceleration $\frac{V^2}{r}$.

Thus to obtain an orbit, the centripetal acceleration $\frac{V^2}{r}$ must equal g

For an altitude of 62 miles (a low earth orbit where there is very little reduction in the gravitational acceleration),

$$g = 31.1206\ \text{ft/sec}^2, r = 2.123028 \times 10^7\ \text{ft},$$

$$V = (31.1206 \times 2.123028 \times 10^7)^{1/2} = 25,704\ \text{ft/sec} = 17,525\ \text{miles/hour}.$$

Referring to Figure 9.1, and from Newton's second law in an inertial frame (the earth), where $F_G - F$ is the accelerating force and F is the reaction force:

$$F_G - F = mg$$

$$mg - F = mg$$

$$F = mg - mg = 0$$

Thus, the reaction force acting on the satellite is zero and it is in free fall (its acceleration is equal to g). Note that this condition of weightlessness is obtained because of the satellite's high tangential velocity and not because of the satellite's height above the earth.

If the forces acting on an object (identified with the subscript 1) in the satellite frame of reference are considered, then from equation 1.2, the force equilibrium equation is $-F_1 + m_1 g - m_1 \frac{V^2}{r} = 0$ where $-m_1 \frac{V^2}{r}$ is the centrifugal force resulting from the satellite's centripetal acceleration and $F_1 = m_1 g - m_1 \frac{V^2}{r} = 0$ where the weight of the object $W_1 = m_1 g$ is balanced by the real centrifugal force $m_1 \frac{V^2}{r}$ and thus

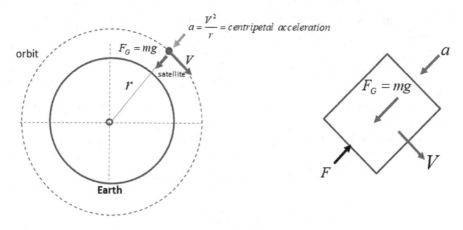

FIGURE 9.1   Satellite orbit and satellite free body diagram in an inertial frame.

the reaction force $F_1 = 0$. Therefore, in the satellite frame of reference, the occupants are weightless and have net zero acceleration relative to the satellite and this is a consequence of free fall in the inertial reference system.

Another solution approach to determine how a satellite maintains an earth orbit is now presented. In this approach, it is stated that if the tangential speed of the satellite is sufficient, then the satellite falls toward the earth at the same rate that the earth curves. This causes the satellite to stay at the same height above the earth and to orbit in a circular path. The combined motion due to the tangential velocity and the fall toward the center of the earth due to the acceleration of gravity (1 g) causes the satellite to fall back into the same circular orbit over and over again. This explanation says that the satellite is falling under the influence of gravity (its acceleration is equal to g) and is in free fall. Referencing Figure 9.2, using the Pythagorean theorem, the following development shows that the free falling explanation does give the correct value for the velocity needed to maintain the orbit.

$$(y+r)^2 = x^2 + r^2$$

$$\left(\frac{g}{2}t^2 + r\right)^2 = V^2 t^2 + r^2$$

$$V^2 = \left(\frac{g}{2}t^2 + r\right)^2 \frac{1}{t^2} - \frac{r^2}{t^2}$$

$$V^2 = \frac{g^2 t^2}{4} + gr + \frac{r^2}{t^2} - \frac{r^2}{t^2} = \frac{g^2 t^2}{4} + gr$$

$$V = \left(\frac{g^2 t^2}{4} + gr\right)^{1/2} = \frac{1}{2}\left(g^2 t^2 + 4gr\right)^{1/2}$$

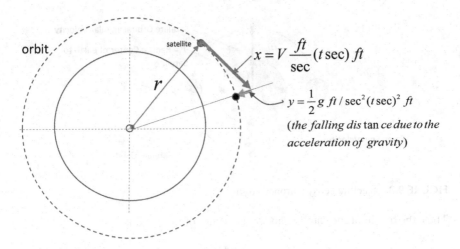

FIGURE 9.2 Satellite tangential and falling distances for the free falling explanation.

Capturing the circular orbit is a continuous process and in the limit as time t approaches zero, the satellite does not leave the circular path and therefore the term g^2t^2 is essentially equal to zero. Also note that there is no tangential force acting on the satellite and thus there is no tangential acceleration as the first term in the above equation for velocity would infer.

Thus,

$$V = (gr)^{1/2} \text{ ft/sec}$$

which is the same value previously developed from the force equation for circular motion (equation 9.1).

Note that the escape velocity for a satellite is the minimum velocity needed to escape from the gravitational influence of the earth. The escape velocity for a satellite in a circular orbit is equal to $\sqrt{2}$ times the velocity needed to maintain the orbit. Thus for an satellite in a circular orbit at an altitude of

$$r = 2.123028 \times 10^7 \text{ ft , the escape velocity is}$$

$$\sqrt{2} \times 17{,}547.3067 \text{ miles/hour} = 24{,}815.637 \text{ miles/hour.}$$

A geosynchronous satellite has an orbit about the earth in which the satellite orbits around the earth with an angular velocity that is equal to the earth's spin speed. The result is that the satellite always remains above the same position relative to the earth all of the time. In Figure 9.3, this position is represented by the black box drawn on the earth and $\omega = \dot{\phi}$.

The period of the earth's spin speed is 86,164.09054 seconds.

From equation 9.1, the orbital speed of the satellite is

$$V = (gr)^{1/2} \text{ ft/sec.}$$

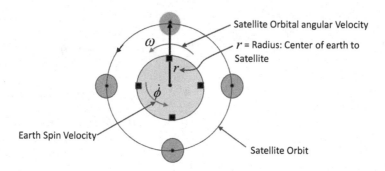

FIGURE 9.3   Satellite geosynchronous orbit.

Thus, the height of the satellite above the center of the earth is

$$r = \frac{V^2}{g}$$

The geosynchronous satellite will be in a high earth orbit and thus the gravitational acceleration g will be significantly reduced. This effect will be explicitly included in the equation for the satellite height r by using the inverse square law:

$$g = \frac{g_0 r_0^2}{r^2},$$

where r_0 is the radius of the earth, g_0 is the surface acceleration ($32.2\,\text{ft/sec}^2$), and g is the local gravitational acceleration at the radius r from the center of the earth.

Thus,

$$r = \frac{V^2}{g} = \frac{V^2 r^2}{g_0 r_0^2}$$

and the satellite orbital period is equal to that of the earth's spin speed, thus

$$V = 2\pi r / T$$

and

$$r = \frac{V^2 r^2}{g_0 r_0^2} = \frac{4\pi^2 r^2}{T^2} \times \frac{r^2}{g_0 r_0^2} = \frac{4\pi^2 r^2}{86{,}164.09054^2} \times \frac{r^2}{g_0 r_0^2}$$

$$r = \sqrt[3]{\frac{86{,}164.09054^2}{4\pi^2} g_0 r_0^2} = \sqrt[3]{\frac{86{,}164.09054^2}{4\pi^2} \times 32 \times \left(2.09022 \times 10^7\right)^2} = 2.0597 \times 10^4\ \text{ft}$$

and

$$r = 2.0597 \times 10^4\ \text{ft} \times 2.8473 \times 10^4\ \text{miles/foot} = 28{,}473\ \text{miles}$$

Subtracting the earth's radius gives an altitude of

$$28{,}473 - 3{,}959 = 24{,}511\ \text{miles} = 21{,}285\ \text{nautical miles}$$

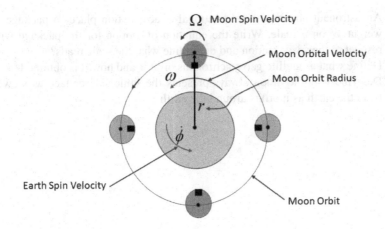

FIGURE 9.4 Moon orbit about the earth.

As discussed in Section 9, the moon is a satellite that is spinning about its axis at an angular velocity equal to that of its orbital angular velocity ($\Omega = \omega$) and this is described as a synchronous orbit of the moon. Figure 9.4 shows the earth and moon as sighted along the earth's polar axis with the black box drawn on the moon representing a reference point that always faces the earth because of the synchronous orbit of the moon. Thus, the moon always presents the same face as viewed from the earth as it orbits around the earth.

The side of the moon not facing the earth is termed the far side of the moon. The far side is colloquially referred to as the "Dark Side", but this name is misleading. As the moon orbits the earth, different parts are in sunlight or darkness at different times and neither side is permanently dark or illuminated.

The moon follows an elliptical orbit path around the earth with an average eccentricity of 0.0549, which means that its orbit is not perfectly circular. Its average orbital distance from the earth is 384,748 km (239,070.43 miles) and ranges from 364,397 km at its closest to 406,731 km at its most distant.

The moon is a big heavy satellite with an appreciable mass compared with the mass of the earth. Consequently, both the moon and earth orbit about their common barycenter (in astronomy, the barycenter is the center of mass of two or more bodies that orbit one another and is the point about which the bodies orbit).

This common barycenter is located inside of the earth, but not at the earth's center. Instead, it lies along the line connecting the earth and moon, roughly two-thirds of the way from the center of the earth to its surface.

PROBLEMS

1. Show that the centripetal acceleration of a satellite in an orbit must equal the gravitational acceleration g.
2. Show that the occupants of a satellite are weightless as a consequence of satellite free fall.

3. An astronaut aboard the International Space Station places a package of weight W on a scale. Write the equation of motion for the package with respect to the space station and determine what the scale reads?
4. Define what a satellite geosynchronous orbit is and how it is obtained.
5. Describe why the moon always presents the same side or face as viewed from the earth as it orbits around the earth.

References

1. Taylor, John R., *Classical Mechanics*, University Science Books, 2005.
2. Anderson, David F. and Eberhardt, Scott, *Understanding Flight*, McGraw Hill, 2001.
3. McLean, Doug, *Understanding Aerodynamics*, Wiley, 2016.
4. Harrison, Hugh, *Aerodynamics for Naval Aviators*, NAVAIR, 1965.
5. Kenny, James, *Principles of Flight*, Oxford Aviation Training, 2017.
6. Sanderson, Jeppesen, *Instrument, Commercial Manual*, Jeppesen Sanderson, 2002.
7. Hunecke, Klaus, *Jet Engines-Fundamentals of the Theory, Design and Operation*, Motorbooks International, 1997.
8. Jeffcott, H.H., "The Lateral Vibration of Loaded Shafts in the Neighborhood of a Whirling Speed - The Effect of Want of Balance", *Phil. Mag*, Series 6, Vol. 37, p. 304, 1919.
9. Logsdon, Tom, *Orbital Mechanics*, Wiley, 1998.
10. Stowell, Rich, *Emergency Maneuver Training*, Rich Stowell Consulting, 1996.
11. Johnston, A.M. "Tex", *Jet-Age Test Pilot*, Smithsonian Institution Press, 1984.
12. Babinsky, Holger, *How Do Wings Work?*, Physics Education, IOP Publishing, 2003.
13. Daso, Dik Alan, *Doolittle Aerospace Visionary*, Potomac Books, 2003.
14. Gregory, James, *The Science of Flight Course Guidebook*, The Great Courses Teaching Company, 2017.
15. Mises, Von, *Theory of Flight*, Dover Publications, 1959.
16. Anderson, John, *Introduction to Flight*, McGraw-Hall, 1978.
17. Federal Aviation Administration, *Pilot's Handbook of Aeronautical Knowledge FAA-H-8083-25B: Flight Training Study Guide*, U.S. Department of Transportation, 2016.

Physics of Flight Problems and Solutions

CHAPTER 1

1. Describe Newton's second law of motion in an inertial frame and with respect to an accelerating frame.

   Newton's second law of motion is usually expressed relative to an inertial frame, and this is called the standard form of Newton's second law. In the standard form, F = force = mass × acceleration, and the force term F is equal to the sum of the forces acting on the mass. In an accelerating frame of reference, Newton's second law is expressed in the nonstandard form whereby in addition to the sum of the forces F acting on the mass, as expressed in the standard form, there is an additional force acting on the mass equal to the mass × the negative of the acceleration of the accelerating frame. Note that in the standard form, the acceleration is relative to the inertial frame and that in the nonstandard form, the acceleration is relative to the accelerating frame.

2. Describe why the inertia force with respect to an accelerating frame is a real force.

   The inertia force relative to the accelerating frame is real force as evidenced by taking a free body diagram of the mass in the accelerating frame. In the accelerating frame, the forces acting on the mass, by definition from Newton's second law, are the sum of the forces equal to F and the inertia force equal to $-mA$.

CHAPTER 2

1. Describe the forces acting on an object with respect to an accelerating frame.

   In an accelerating frame, the forces acting on a mass are the resultant of all forces plus the inertia force $-mA$, where A is the acceleration of the accelerating frame.

2. Describe the equivalence of Newton's second law of motion in inertial and accelerating frames when the object in the accelerating frame has zero acceleration with respect to the accelerating frame.

   The nonstandard and standard versions of Newton's second law of motion have exactly the same form, except that in addition to the sum of the forces F, the nonstandard form has an extra force $-mA$, termed the inertia force.

When the mass is at rest in the accelerating frame, that is, its acceleration relative to the accelerating frame is zero, then $F = mA$ is the same equation of motion relative to both the inertia and accelerating frames.

CHAPTER 3

1. Describe Newton's second law of motion with respect to a rotating frame.
   For a mass at rest in a rotating frame, then relative to the rotating fame, the nonstandard form of Newton's second law of motion is $0 = m\ddot{r} = F - mA = F - m\dfrac{V^2}{r}$ where A is the centripetal acceleration and V is the tangential velocity of a mass traveling in a circular path with radius r relative to an inertial frame. The inertial force $-mA = -m\dfrac{V^2}{r}$ is an outward or centrifugal force relative to the rotating frame and since the mass has no velocity relative to the rotating frame, that is, the Coriolis force is zero, and thus, the acceleration $\ddot{r}$ relative to the rotating frame is zero.

2. Describe why in a rotating frame that the centrifugal force is a real force.
   The nonstandard form of Newton's second law of motion shows that in a rotating frame of reference, the centipedal acceleration A relative to an inertial frame results in an inertia (centrifugal) force $-mA = -m\dfrac{V^2}{r}$ that is real relative to the rotating frame.

CHAPTER 4

1. Describe what free fall is and why it results in a weightless condition.
   When an object is in free fall, its acceleration is equal to the acceleration of gravity as the only force acting on the object is the gravity force. Newton's second law in an inertial frame shows that the object is weightless because the object's weight is balanced by the product of its mass and the acceleration of gravity.

2. Develop the equation that defines the aircraft velocity needed to achieve free fall at the top of the maneuver used for the so-called reduced gravity aircraft.
   When the aircraft is in free fall in the upper zone of the parabolic flight path, Newton's second law provides $W = mg = m\dfrac{v^2}{r_p}$. Thus, the aircraft's required velocity is $v = \sqrt{gr_p}$ where r_p is the instantaneous radius of curvature.

3. What are the forces acting on the payload of the so-called reduced gravity aircraft when it is in free fall?
   The forces acting on the payload relative to the aircraft are the centrifugal force resulting from the aircraft's downward acceleration of gravity and the gravity force, resulting in net zero force. Thus, the payload is weightless.

CHAPTER 5

1. Explain why the use of Bernoulli's theorem to describe how the lift force is created is clearly incorrect.

   The explanation of how lift is developed using Bernoulli's equation is based on the equal transit time principle which states that the airflow separated at the leading edge of the airfoil must rejoin at the trailing edge at the same time. However, this is oeever clearly incorrect as shown by wind-tunnel experiments.

2. Describe the alternative explanation of lift based on airflow curvature.

   It is theorized that the conservation of momentum explicitly expressed by Newton's second law provides the basis for understanding how aerody-namic lift is developed. Airfoil curvature compels curvature of the stream-lines resulting in normal to the streamline accelerations in the fluid field. These accelerations are sustained by normal pressure forces acting above and below the airfoil surfaces, and the net result is the lift force.

3. Describe the effect of angle of attack on the lift force and on the stall speed.

   The angle of attack is the angle between the chord line of the airfoil and the rela-tive wind. The lift coefficient is a dimensionless parameter that is a function of the angle of attack and expresses the lift force as a function of the fluid velocity, the fluid density, and an associated reference area. The lift coefficient increases with increasing angle of attack until the critical angle of attack is reached, caus-ing a reduction in the lift coefficient and thus a loss of lift as the airfoil stalls.

4. Describe the effect of the load factor on the stall speed.

   The lift force is equal to the aircraft's weight times the load factor and the effect of an increased load factor is to increase the required velocity for a given angle of attack, and thus, the velocity at stall is proportional to the square root of the load factor.

CHAPTER 6

1. What unbalanced force causes an airplane to turn?

   When the airplane is banked, an unopposed horizontal lift force compo-nent is developed that radially accelerates the airplane causing the center of gravity of the airplane to follow a circular path.

2. What is the airplane primary turn control?

   The ailerons act as the primary turn control by banking the airplane to pro-duce an unopposed horizontal lift force component that causes the airplane to follow a circular path.

3. What are the forces acting on an airplane in both coordinated and uncoor-dinated level turns?

   The forces acting on an airplane in both coordinated and uncoordinated level turns are the airplane's weight and the airplane's resultant aerodynamic lift force. Note that the centrifugal force is often identified as an additional force. However, in an inertial reference system, the centrifugal force which is equal to the product of the radial acceleration and the mass of the airplane is not an actual force. However, in the airplane's reference frame, the centrifugal force is an actual force.

4. What is an airplane slip or skid, what causes this behavior in uncoordinated turns, and what can be done to avoid these phenomena?

   When an airplane is slipping or skidding in a turn, it is undergoing an uncoordinated turn such that the airplane's longitudinal axis is not perpendicular to the turn radius. Note that in this condition, the airplane's resultant aerodynamic lift force does not lie in the plane of symmetry perpendicular to the airplane's lateral axis. Also note contrary to popular belief, that there must be equilibrium between the horizontal lift force component and the centrifugal force in the slipping or skidding turns. The slipping or skidding turn is avoided by proper coordination between the rudder and the ailerons controls to maintain the airplane's longitudinal axis tangent to the turn circle.

5. Describe the effect of bank angle on the stall speed.

   The stall speed increases with increasing bank angle in a turn because the resulting increase in load factor n increases the effective weight of the aircraft.

6. Describe the velocity-load factor diagram, its purpose, and how it is generated.

   The Velocity-Load Factor diagram (V-n diagram) relates the airspeed V at stall to the load factor (n). It is specific to an airplane's configuration and altitude and provides the maximum amount of positive or negative lift the airplane is capable of generating at a given speed. It includes the safe load factor limits and the safe load factor the airplane can sustain at various speeds. The load factors for the maximum positive and negative lift curves are generated for speeds at the critical angles of attack to construct the V-n diagram.

7. Define maneuver speed.

   Maneuver speed V_a is the speed at which an airplane will stall before exceeding its maximum limit load. It is defined as the intersection of the positive limit load factor line and the maximum positive lift curve in the Velocity-Load factor diagram (point B in Figure 6.3.1). Also note from equation 5.2.3, that as the airplane's weight is increased, the maneuver speed increases.

8. When can the condition of free fall and thus weightlessness occur in a loop and what is the governing equation defining the airplane's velocity for this condition?

   The condition of free fall and thus weightlessness can occur in a loop. For example, from equation 6.4.1, the lift L at the top of an inside loop is equal to $m\left(\dfrac{v^2}{r} - g\right)$ and if the lift force is equal to zero, then $m\dfrac{v^2}{r} = mg$ and thus

$\dfrac{v^2}{r} = g$. Therefore, the only force acting on the airplane is the gravity force mg and the airplane's acceleration is equal to the acceleration of gravity. Thus, the airplane is in free fall and weightless. The airplane's velocity for this condition is given by $v = \sqrt{gr}$.

9. Why does the pilot experience negative gs when an aileron roll is performed?
   When the aileron roll is performed, the airplane is rotated about its longi-tudinal axis through the $1g$ field such that relative to the pilot, the $1g$ vector is rotating.

   Hence, at the initiation of the roll, the $1g$ vector is pointed head to foot, providing $+1g$, and $180°$ later, the $1g$ vector is pointed foot to head resulting in the pilot experiencing negative $1g$.

10. Describe why a uniform load factor of $+1$ cannot be obtained in a barrel roll.
    Because of the combination of gravitational and centrifugal acceleration forces experienced during the barrel roll, a uniform load factor of $+1$ cannot be obtained. This can be seen from inspection of the load factor equation:

$$n = \sqrt{(\sin\theta)^2 + \left(\frac{r\dot{\theta}^2}{g} + \cos\theta\right)^2}$$

where θ is the roll angle.

11. Explain why counter rotating propellers were used in the Wright Brothers airplanes.
    The gyroscopic moments acting on the airplane produced by the rotating propellers are cancelled when the propellers are counter rotating.

CHAPTER 7

1. Explain how Newton's second law is used to describe how thrust is pro-duced for all aircraft propulsion systems.
   The thrust force F is essentially dependent on the rate of change of the momentum of the working fluid (air), as described by Newton's second law of motion:

$$F = ma = \frac{dmv}{dt} = m\frac{dv}{dt} = \dot{m}(V_e - V_0).$$

where $\dot{m}$ is the mass flow rate of the working fluid. V_e and V_0 are the working fluid exit and entry velocities, respectively.
   This is true for all aircraft propulsion systems.

2. Explain why variable pitch propellers are more efficient in producing thrust and define what propeller blade feathering is and how it is used to reduce propeller drag in the event of engine failure.

   Fixed pitch propellers are efficient only at a given combination of airspeed and engine power. In order to provide the blade pitch angle β to obtain the most efficient angle of attack over different airspeed and engine power conditions, variable pitch propellers are used whereby the blade can be rotated around its long axis to change the blade pitch. In the event of engine failure, the propeller control mechanism for variable pitch propellers allows the propeller to be feathered to reduce propeller drag. This is accomplished by changing the pitch angle to 90° such that the propeller chord is parallel to the direction of flight and thus the propeller stops rotating and minimum windmilling, if any occurs.

3. Explain the difference between turbojet and turbofan engines and why turbofan engines are more efficient.

   From Newton's second law of motion, jet engine thrust is essentially described by the following equation.

$$F = ma = \frac{dmv}{dt} = m\frac{dv}{dt} = \dot{m}(V_e - V_0)$$

   This equation shows that thrust is dependent on both the air mass flow rate and the air exit velocity. In a turbojet engine, the acceleration of the airflow through the engine core produces all of the thrust where the air exit velocity dominates over the mass flow rate in producing the thrust. Turbofans are able to accelerate significant amounts of air without burning additional fuel because of the fan. The fan extracts some additional energy from the exhaust via the turbine, slowing the exhaust velocity slightly, but substantially adds to the mass flow rate via the bypass air. This results in high propulsive efficiency (the work done to propel the aircraft divided by the work done by the engine to accelerate the engine's air mass).

CHAPTER 8

1. Provide the definition of the critical speed for piston, jet, and gas turbine engines.

   In piston, gas turbine, and jet engines, the amplitude of vibration of the engine system often passes through a maximum at a speed that is called the critical speed. The critical speed is the rotor spin speed (rotational speed), which is equal to the engine system natural frequency. The amplitude of vibration is commonly excited by rotor unbalance.

2. Show how the use of Newton's second law provides the dynamic equations of motion that address the effects of the rotating components in aero engines.

Newton's second law is used to construct the equations of motion that relate the system dynamic response to the forced response associated with rotor unbalance and the excitation frequency (rotor spin speed). The system dynamic equations of motion include the stiffness, damping, and rotor unbalance forces, and address the phase shift of the rotor's center of mass as the critical speed is traversed.

3. Discuss why engine lateral synchronous vibration response due to rotor unbalance is an important consideration in engine design.
   The forced response of the engine system due to rotor unbalance occurs at a frequency equal to the spin speed of the rotor and this behavior is termed lateral synchronous vibration. This results in dynamic stresses in the static structures and static stresses in the rotor. Important considerations are the rotor-static structures relative dynamic displacements compared to the running clearances, and the allowable stresses compared to the response stresses. Hence, the operability and structural integrity of the engine is dependent on the lateral synchronous vibration response due to rotor unbalance.

4. Discuss the response behavior of the engine rotor when it is spinning at speeds below the critical speed, at the critical speed, and above the critical speed.
   When the rotor is spinning at speeds below the critical speed, the heavy side is outside the rotor orbit. When the rotor operating at a spin speed equal to the critical speed, the unbalanced force leads the rotor displacement vector by 90°. For spin speeds significantly above the critical speed, the rotor mass center moves to the undeflected center of the rotor such that the heavy side is inside of the orbit and the geometric center of the rotor is spinning and whirling about the mass center.

CHAPTER 9

1. Show that the centripetal acceleration of a satellite in orbit must equal the gravitational acceleration g.
   The gravitational force acting on a satellite is $F_G = mg$ and this is equal to its centripetal acceleration × the mass m of the satellite:

$$mg = m\frac{V^2}{r}$$

$$g = \frac{V^2}{r}$$

r is the distance between the satellite and the center of the earth.
   Thus, the centripetal acceleration of a satellite in orbit is equal to the gravitational acceleration g.

2. Show that the occupants of a satellite are weightless as a consequence of satellite free fall.

The satellite is in free fall since the only force acting on a satellite is the force of gravity, and its acceleration is equal to the acceleration of gravity. As a consequence, the centrifugal force mg acting on an occupant of the satellite that results from the satellite's centripetal acceleration is equal and opposite to the weight mg of the occupant. Therefore, the occupant is weightless.

3. An astronaut aboard the International Space Station places a package of weight W on a scale. Write the equation of motion for the package with respect to the space station and determine what the scale reads?

The package weight is equal to $W = m\,g$ and the package inertial force due to the satellite's centripetal acceleration is $-mg$. The force exerted on the package by the scale is $-F$. Note that outward directed forces have been assigned negative signs.

$$-mg - F + m\,g = 0$$

$$F = -mg + mg = 0$$

Thus, the force exerted on the package by the scale is zero and therefore the scale reads zero since the package is weightless.

4. Define what a satellite geosynchronous orbit is and how it is obtained.

A geosynchronous satellite orbits around the earth with an angular velocity that is equal to the earth's spin speed. The result is that the satellite always remains above the same position relative to the earth all of the time. Geosynchronous satellites are in high earth orbit where the gravitational acceleration g is significantly reduced.

5. Describe why the moon always presents the same side or face as viewed from the earth as it orbits round the earth.

The moon is a satellite that is spinning about its axis at an angular velocity equal to its orbital angular velocity around the earth. This is described as a synchronous orbit of the moon. Thus, the moon always presents the same face as viewed from the earth as it orbits around the earth.

Index

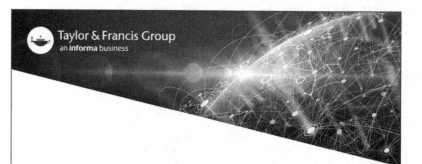

Printed in the United States
by Baker & Taylor Publisher Services